INTRODUCING
Newton

William Rankin

Edited by Richard Appignanesi

D0301041

Icon Books UK Totem Books USA

This edition published in the UK in 2007 by Icon Books Ltd., The Old Dairy, Brook Road, Thriplow, Cambridge SG8 7RG email: info@iconbooks.co.uk www.introducingbooks.com

Sold in the UK, Europe, South Africa and Asia by Faber and Faber Ltd., 3 Queen Square, London WC1N 3AU or their agents

Distributed in the UK, Europe, South Africa and Asia by TBS Ltd., TBS Distribution Centre, Colchester Road, Frating Green, Colchester CO7 7DW

This edition published in Australia in 2007 by Allen & Unwin Pty. Ltd., PO Box 8500, 83 Alexander Street, Crows Nest, NSW 2065

Previously published in the UK and Australia in 1993 as *Newton for Beginners* and in 2000 as *Introducing Newton and Classical Physics*

Reprinted 1994, 1995, 1997

This edition published in the USA in 2007 by Totem Books Inquiries to Icon Books Ltd., The Old Dairy, Brook Road, Thriplow, Cambridge SG8 7RG, UK

Previously published in the USA in 1994 as *Introducing Newton* and in 2000 as *Introducing Newton and Classical Physics*

Distributed to the trade in the USA by National Book Network Inc., 4501 Forbes Boulevard, Suite 200, Lanham, Maryland 20706

Distributed in Canada by Penguin Books Canada, 90 Eglinton Avenue East, Suite 700, Toronto, Ontario M4P 2YE

ISBN-10: 1-84046-842-4
ISBN-13: 978-1840468-42-7

Originating editor: Richard Appignanesi

Printed by Gutenberg Press, Malta

NEWTON

Qui genus humanum ingenio superavit.

"Who surpassed all men in genius"

There is a statue of Isaac Newton in the antechapel of Trinity College, Cambridge. The poet Wordsworth on seeing it bathed in moonlight from his pillow, wrote...

"... Newton, with his prism and silent face,
The marble index of a mind for ever
Voyaging through strange seas of Thought,
alone."

By Wordsworth's time Newton was already completely transformed from a creature of flesh and blood to a stone-faced demi-god presiding over the Industrial Revolution.

"One continued series of labour, patience, humility, temperance, meekness, humanity, beneficence and piety without any tincture of vice." — John Conduitt.

It is a long way from the retiring schoolboy who carved his name in a windowsill to the man who set his indelible mark on the centuries that followed. We must travel back to the beginnings of civilization to find the wellsprings of the scientific revolution that would change the world.

It's the Thought that Counts

Our story has its beginnings in the simple practical activities of everyday life.

On the Banks of the Nile

The Egyptians had an intimate relation with the heavens and depended on the seasonal flooding of the Nile to bring fertility to their fields. These fields were taxed according to area. Each year it was necessary to check to see if land had been swept away to set the correct tax.

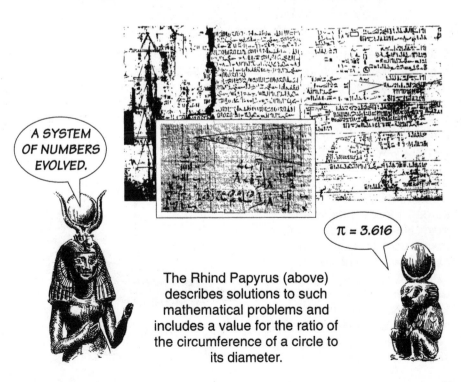

The Rhind Papyrus (above) describes solutions to such mathematical problems and includes a value for the ratio of the circumference of a circle to its diameter.

By the Waters of Babylon

In the fertile lands irrigated by the Tigris and Euphrates a civilisation grew which recorded the movements of the heavens over thousands of years. The Babylonians had a system of numbers based on a unit of 60 which permitted calculations with very large numbers. Traces of their system still remain with 60 seconds making a minute, and 360 degrees to a circle.

Babylonian reckoning was certainly more advanced than that of Egypt but it was still only a collection of prescription-like rules for calculating areas with no proof. There was no logical method to apply to new problems as they turned up.

For a deductive system based on proofs, we have to await the return to the Greek island of Samos of a man who had been wandering thirty-four years abroad 'mongst priests and magi. He transformed number from a useful tool into the central principle of life. He called his new philosophy μαθηματικη (mathematics). Eight hundred left their homes and families to follow him when he first presented his ideas as a sermon on a mount.

All is Number

HE STUDIED THE MYSTERIES IN EGYPT.

AND AMONGST THE CHALDEANS.

HE IS THE SON OF THE GOD APOLLO.

NO, HIS FATHER IS THAT SUBSTANTIAL CITIZEN MNESARCHAS.

THERE ARE MEN, THERE ARE GODS, AND THERE ARE BEINGS LIKE ME.

Pythagoras was a combination of Einstein and the Maharishi. He advocated a religion based on the transmigration of souls and the sinfulness of eating beans. He preached to the animals.

ALL THINGS BORN WITH LIFE IN THEM OUGHT TO BE TREATED AS EQUALS.

Pythagoras, 572 - 480 B.C.

THE WHOLE HEAVEN IS NUMBER AND HARMONY.

INCLUDING WOMEN.

WHAT ABOUT CATS?

Pythagoras discovered the connection between number and music: that the pitch of a note depends on the length of the string that produces it.

The sounds made by the planets as they sped through space combined to produce a music, "The Harmony of the Spheres". This harmony was soon disturbed from within.

In the society he founded, men and women were equal, property was held in common, and even mathematical discoveries were collective.

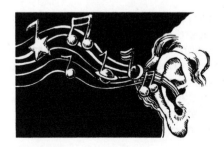

9

A Cloud of Infinity

That we still call numbers odd or even or talk of squares and cubes of numbers is due to Pythagoras. But he is best known for the *Pythagoras Theorem*. It was to destroy his order.

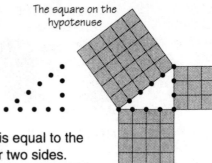

The square on the hypotenuse

The square on the hypotenuse is equal to the sum of the squares on the other two sides.

3^2 (3 squared) + 4^2 (4 squared) = 9 + 16 = 25

The hypotenuse = $\sqrt{25}$ (the square root of 25)

So the hypotenuse = 5

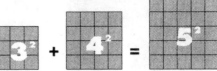

A pythagorean called Hippasus on a boat trip thought that finding the diagonal of a square would be a harmless pastime.

1^2 (one squared) + 1^2 (one squared) = 1 + 1 = 2

The diagonal = $\sqrt{2}$ (the square root of 2)

But what is the square root of two?

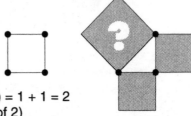

All attempts to express root two as a fraction (a ratio of two whole numbers) failed. There is no such ratio, root two is irrational. Here was something which had to be a number as it had a length, and yet it could not be written down.

Hippasus was thrown overboard and the Brotherhood sworn to secrecy, but the harm was done. All is number, but not all numbers are numbers. Odd and even at the same time, the Irrational destroyed the Harmony of the Spheres.

AN IRRATIONAL NUMBER IS NOT A TRUE NUMBER BUT LIES HIDDEN IN A CLOUD OF INFINITY.

Michael Stifel in
Arithmetica Intigra, 1544

Squaring the Circle

WHY WOULD ANYBODY WANT TO CHANGE A CIRCLE TO A SQUARE?

USING ONLY A RULER AND COMPASSES.

Irrationals were evaded by treating all numbers as lengths, but the square root of two wasn't the only problem the Greeks had. The cream of Hellenic intellect was tied up for hundreds of years trying to square the circle.

The problem comes down to determining the ratio of the circumference of a circle to its diameter. This ratio is called π.

Despite the repeated efforts of the best Greek mathematicians no circle was ever squared, nor was it for the next 2,000 years. One hundred years ago it was finally proved to be impossible.

WHAT WILL I GAIN BY STUDYING GEOMETRY?

SLAVE, GIVE THE BOY A PENNY, SINCE HE MUST PROFIT BY HIS LEARNING.

Euclid, 300 B.C.

The Greeks were scornful of utility, and indeed Plato thought that the degrading trade of shopkeeping should be punishable as a crime. Typically they spared no effort pondering impossible problems. One of the by-products was a series of useless curves, those produced by slicing up cones at different angles.

The Conic Sections

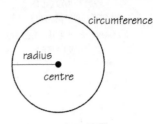

Circle
Bounded by a line (circumference) which is everywhere the same distance (radius) from a fixed point (centre).

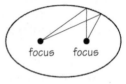

Ellipse
A curve traced by a point which moves so that the sum of its distances from two fixed points (foci) is constant.

Parabola
A curve traced by a point which moves so that its distance from a fixed point (focus) is equal to its distance from a given straight line (directrix).

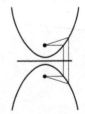

Hyperbola
A curve traced by a point which moves so that its distance from a fixed point always has a value greater than one to its distance from a fixed line (directrix).

Exhaustion

Talking of side effects, more pondering threw up an even more terrifying problem; the infinite.

First inscribe a triangle inside the circle

Antiphon the Sophist (ca. 430 B.C.) was trying to determine the area of a circle by filling it with triangles. He could then add up the areas of the triangles to get the area of the circle. First he inscribed a triangle. Then he filled the spaces left over with an ever-increasing number of smaller and smaller triangles until the area was "exhausted". There was only one problem.

HE JUST DOESN'T KNOW WHEN TO STOP.

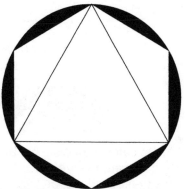

Then fill the spaces left over with smaller triangles until the area is exhausted.

NOTHING THAT IS VAST ENTERS THE LIFE OF MORTALS WITHOUT A CURSE.

Plus Ça Change

Zeno, a follower of Parmenides and his doctrine of the One, set out to prove the non-existence of the Many. He lost his head for treason but first he proposed a series of immortal paradoxes. One of them was *The Achilles*.

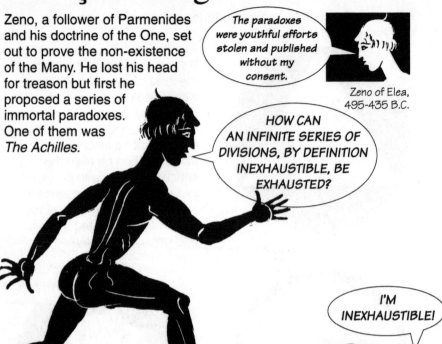

The paradoxes were youthful efforts stolen and published without my consent.

Zeno of Elea, 495-435 B.C.

HOW CAN AN INFINITE SERIES OF DIVISIONS, BY DEFINITION INEXHAUSTIBLE, BE EXHAUSTED?

I'M INEXHAUSTIBLE!

Achilles, the fastest runner of his day, tries to catch a tortoise, but by the time he reaches the place from which it started, the tortoise has moved on. And in the time it takes Achilles to reach that spot, the tortoise has moved on again, and so on ad infinitum. The distance separating Achilles from the tortoise will continually diminish but will never completely vanish. The slower will never be overtaken by the quicker.

AT EVERY INSTANT OF FLIGHT I AM AT REST.

INFINITE MAGNITUDES ASTONISH OUR BRAIN, WHICH IS ABOUT 6" LONG, 5" WIDE AND 6" DEEP, IN THE LARGEST HEADS.

The Achilles and Zeno's other paradoxes confounded the Greeks, who were paralysed by the "Horror of the Infinite" until Archimedes came to the rescue.

Voltaire, 1694-1778

Skirting the Problem

Archimedes, 287-212 B.C.

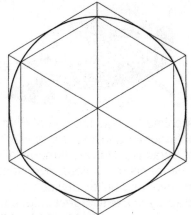

Exhaustion starts
with a polygon inside
the circle.

Archimedes is best remembered for dashing naked through the streets when he discovered his laws of floating bodies.

He skirted the problem of the infinite not by using infinitely small numbers, but by stopping when the number became "as small as we please".

He combined exhaustion with compression, and by successively doubling the number of sides of the polygons to 96, he calculated the value of π to be less than 3 1/7 and greater than 3 10/71.

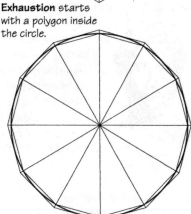

Compression puts
a polygon outside
the circle.

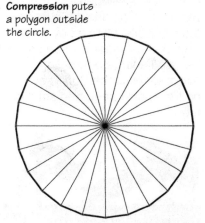

15

Geometry at Play

GIVE ME A PLACE TO STAND AND I CAN MOVE THE WORLD.

Archimedes was the son of Phidias the astronomer and intimate of Hieron, King of Syracuse. He was famous in his own time for his marvellous inventions such as lever and pulley devices. He could drag Syracuse's largest ship, fully loaded, over land with one hand.

Plutarch

"Yet he would not deign to leave behind him any written work on these subjects, but, regarding as ignoble and sordid the business of mechanics and every form of art which is directed to use and profit, he placed his whole ambition in those speculations, the beauty and subtlety of which is untainted by any admixture of the common needs of life."

In 212 B.C. the Roman army under the command of Marcellus lay siege to Syracuse. Romans sat outside the city walls, stymied by Archimedes' ingenious engines. He used a large parabolic mirror to focus the Sun's rays on the enemy fleet, and set it on fire.

Marcellus

BY THE MULTITUDE OF THE MISSILES THAT HE HURLS AT US ALL AT ONCE HE OUTDOES THE HUNDRED-HANDED GIANTS OF MYTHOLOGY.

MERELY THE DIVERSIONS OF GEOMETRY AT PLAY.

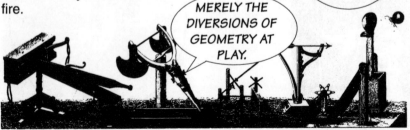

Death of Original Thought

The Roman fleet pretended to withdraw, but it returned secretly and breached the wall during a festival for the goddess Diana. Syracuse was given entirely that day to wine and sport.

"Archimedes would often forget his food and neglect his person, to the degree that when he was occasionally carried by absolute violence to bathe or have his body anointed, he used to trace geometrical figures in the ashes of the fire, and diagrams in the oil on his body, being in a state of entire preoccupation" — Plutarch.

Unaware that the city had been taken, Archimedes was engrossed when a shadow fell across his diagram.

Famous last words

DON'T DISTURB MY CIRCLES.

This is the only appearance of a Roman in the history of mathematics. The theoretical Greeks, with their love of abstract science, were superseded in the leadership of Europe by the practical Romans.

The Roman soldier who killed Archimedes is a symbol of the death of original thought that Rome caused throughout the Hellenic world

"No Roman ever lost his head because he was absorbed in the contemplation of a mathematical diagram." — Whitehead

Bertrand Russell

Nature and Nature's laws,
lay hid in night:
God said, *Let Newton be!*
and all was light.

Alexander Pope, 1688 - 1744

Born so Little

It was just past midnight, Christmas Day, 1642, when Isaac Newton was born prematurely at the Manor of Woolsthorpe in Lincolnshire.

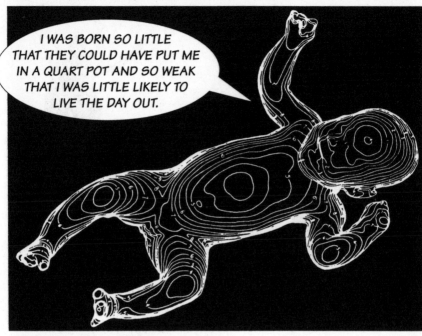

I WAS BORN SO LITTLE THAT THEY COULD HAVE PUT ME IN A QUART POT AND SO WEAK THAT I WAS LITTLE LIKELY TO LIVE THE DAY OUT.

It was the year of that argumentative Italian physicist Galileo's death.

NO IT WASN'T!

In go-ahead Italy it was already January the 4th 1643! The Pope had recently introduced a new, more accurate calendar. But the English were having none of it.

WE BRITS PREFER TO DISAGREE WITH THE SUN THAN TO AGREE WITH THE POPE.

ANY DAY!

What Day is It?

The problem was that a year of the Julian calendar was 11 minutes 12 seconds too long. By the 1500s the Spring Equinox was ten days out, and Easter was well off course and heading towards summer.

A Neapolitan doctor had the answer.

Introduced by me, and July is my month.

Julius Caesar, 100-44 B.C.

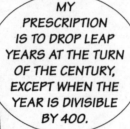

MY PRESCRIPTION IS TO DROP LEAP YEARS AT THE TURN OF THE CENTURY, EXCEPT WHEN THE YEAR IS DIVISIBLE BY 400.

Luigi Lilio, d. 1576

RIGHT! IT'S MY CALENDAR, I'M TAKING TEN DAYS OFF THIS YEAR RIGHT AWAY.

Pope Gregory XIII

They don't make calendars like they used to!

No Pope's taking days off my life!

Popish trickery!

If it was good enough for Julius Caesar, it's good enough for me!

The new "Gregorian" calendar wasn't accepted in Germany, Denmark and Norway until 1700, in England and Sweden 1743, Japan 1873, China 1912, Russia 1918, and Greece 1923.

Childe Isaac

Newton's infancy was blighted. His father, Isaac, "a wild, extravagant and weak" yeoman farmer, had died before the birth.

When Newton was three, his mother, Hannah, remarried. Her new husband was the 63-year-old Rector of North Witham, Barnabas Smith, and she moved to his parish.

I WILL BURNE YOU AND THE HOUSE OVER YOU!

Isaac was left at Woolsthorpe to be raised by his maternal grandmother. If his father had lived, he might never have received an education.

I see your Uncle Bill has been made Rector of Burton Coggles.

Fine family we Ayscoughs, no Newton has ever been able to write his own name.

WHEN I GROW UP I'LL SHOW 'EM!

Isaac was 10 when Smith died.

I'D BETTER BUILD A BOOKSHELF.

HI ISAAC! I'M HOME.

Hannah, not yet 1 yr.

Benjamin, 3 yrs.

Mary, 6 yrs.

Skool

Aged 12, Newton was sent to school in Grantham. The Free Grammar School of King Edward VI had been founded in the 14th century. Grammar schools were so called because that's mostly what they taught, Latin grammar.

Latin, bit of Greek, Latin, the Vulgate, more Latin and no Maths..

THIS MIGHT COME IN HANDY.

Latin is a language...

... dead as dead can be...

... first it killed the Romans...

... now it's killing me.

Isaac proved to be able to catch up on Maths himself, but without Latin he would have been permanently handicapped. Latin was the international language of European scholarship; in it were written all the important works. Newton's command of Latin, which he came to read and write as fluently as English, enabled him to absorb these books, and when the time came, to communicate his own discoveries back to Europe.

Windmills

At school Newton would leap ahead whenever he put his mind to his studies. But he would often neglect them for strange inventions and showed "an extraordinary inclination for mechanical works", even on the Sabbath.

I KNOW I SHOULDN'T, BUT...

"He had got little saws, hatchets, hammers and a whole shop of tools, which he would use with great dexterity."

Nature and Art

Two books inspired him, and would permanently affect his whole life. The first was John Bate's *The Mysteries of Nature and Art*. The approach it fostered — practical experiment, craftsmanship, chemistry, analysis, organizing into categories — stayed with Newton for the rest of his life.

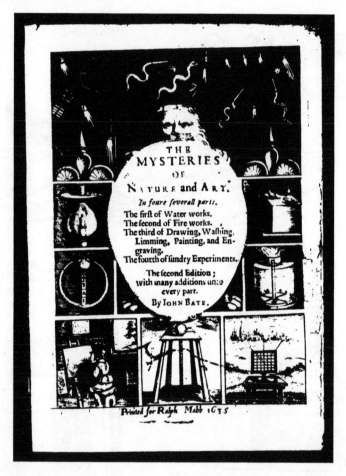

The second was empty, it was a notebook bought for twopence halfpenny. Newton entered notes from Bate at one end, and at the other made alphabetical lists of words under various headings: Artes, Birdes, Cloathes &c. Such careful organizing and categorizing of information would become the mature Newton's hallmark.
This notebook, the first of thousands, described among other things how to make a sundial.

What's the Time?

Behind the organ in Colsterworth church, set in the wall, is a stone.

"Newton: aged 9 years cut with his penknife this dial."

At Grantham, Newton lodged with Mr. Clarke, the Apothecary. He filled the house with sundials.

"In the yard of the house he would drive pegs to mark the hours and half hours made by the shade, which by degrees from some years of observations, he had made very exact, and anybody knew what time it was by Isaac's dials."

Newton's fascination with the Sun's movement never left him. In old age, if asked the time he would consult a shadow instead of looking at the clock.

Fireworks

80   *The second Booke*

thereof; then hang the wings on in such wise, that they may shake as the Dragon runnes along the line; you may dispose divers small serpents in the wings; marke the figure.

How to make fire Drakes.

YOu must take a peece of linnen cloth of a yard or more in length; it must bee cut after the forme of a pane of glasse; fasten two light stickes crosse the same, to

make it stand at breadth; then smeare it over with linseed oyle, and liquid varnish tempered together, or else wet it with oyle of peter, and unto the longest corner fasten a match

You muft take a peece of linnen cloth of a yard or more in length; it must bee cut after the form of a pane of glaffe; fasten two light fticks croffe the fame, to make it ftand at breadth; then fmeare it over with linfeed oyle, and liquid varnifh tempered together, or elfe wet it with oyle of peter, and unto the longeft corner faften a match prepared with faltpeter water upon which you may faften diverfe crackers, or fquibs; betwixt every of which, binde a knot of paper fhavings, which will make it fly the better; then tie a thin rope of fufficient length to raife it unto what height you fhall defire and guide it withall...

Strange Inventions

...then fire the match, and raife it againft the wind in an open field; and as the match burneth it will fire the crackers, and fquibs, which will give diverfe blows in the ayre; and when the fire is once come unto the ftoupell*, that will fire the cloth, which will fhow very ftrangely and fearfully.

*ftoupell = fuse

When darkness fell Isaac would sneak out and fly his exploding kite over the village...

"... wonderfully affrighting all the neighbouring inhabitants for some time, and causing not a little discourse on market days, among the country people, when over their mugs of ale."

AND I SAY IT'S EXTRA-TERRESTRIALS.

IT'LL BE INEXPLICABLE CROP CIRCLES NEXT.

IT'S THEM SHEEP UP TO NO GOOD.

Sober Silent Thinking Lad

The apothecary's stepdaughter, Miss Storer, claimed in later life to have had a romance with the teenage Newton, who made dolls' furniture for her and her friends.

HE WAS A SOBER SILENT THINKING LAD, NEVER KNOWN SCARCE TO PLAY WITH BOYS ABROAD.

He was no sissy, however. She particularly remembered one dispute between her brother Arthur and young Isaac.

"Tho' not so lusty as his antagonist, he had so much more spirit and resolution that he beat him. Isaac pulled him along by the ears & thrust his face against the side of the church, to use him like a coward and rub his nose against the wall."

PERHAPS I SHOULD INVENT THE SPRAY CAN.

Extreme thoroughness in the prosecution of a dispute would also be an enduring characteristic.
But his usual behaviour when faced by a wall was to draw on it. Birds, men, ships, plants, John Donne, schoolmaster Stokes, circles and triangles, King Charles I.

First Experiment

BUT THERE IS NO KING

Oliver Cromwell, 1599-1658

Isaac Newton's childhood was spent under a military dictatorship — that of Cromwell's New Model Army of Puritans known as Roundheads. They had fought for the right of Parliament rather than the monarchy to hold supreme power in England. But having defeated the royalist Cavaliers in civil war, Cromwell dissolved Parliament!

Take away that fool's bauble, the mace.

Not a dog barked.

IT'S A FOOT STRONGER THAN EVER BEFORE.

On the day of Cromwell's death a great storm swept England. The country folk said the Devil was riding the storm to come for Cromwell's soul. Isaac jumped, but not for joy. He was measuring the force of the storm by variously leaping with the wind and against it.

"His school fellows generally were not very affectionate toward him. He was commonly too cunning for them in everything. He who has most understanding is least regarded."

What was worse, his mother took him away from school altogether.

Low Employments

When Isaac was 17, the family tried to make a farmer of him.

On market days he bribed his servant to run things and retired to the Apothecary's house where there was a great parcel of books. The books were left by Dr. Clarke the apothecary's brother, a student of Henry More at Trinity College, Cambridge.

He even ran up a criminal record.

Teacher's Pet

Isaac was increasingly absent-minded, peevish, and argumentative. Finally, his uncle William and Stokes the schoolmaster persuaded Isaac's mother to send him back to school in Grantham to prepare for University.

On his last day at school he was presented as a shining example to follow.

HIS GENIUS NOW BEGINS TO MOUNT UPWARDS APACE & SHINE OUT WITH MORE STRENGTH. HE EXCELS PARTICULARLY IN MAKING VERSES. IN EVERYTHING HE UNDERTAKES HE DISCOVERS AN APPLICATION EQUAL TO THE PREGNANCY OF HIS PARTS & EXCEEDS EVEN THE MOST SANGUINE EXPECTATIONS I HAVE CONCEIVED OF HIM.

Restoration

During Isaac's last months in Lincolnshire, church bells had proclaimed the Restoration of the Monarchy. Republican politics had died with Cromwell. A bargain was struck between the town merchants and the landowners, and the monarchy was restored in the shape of Charles I's son, who became Charles II. But royal power was now severely limited. For instance, the King could no longer impose his own taxes or order arbitrary arrests.

Charles I, King by divine right.

Charles II, by kind permission of landlords and merchants.

Most people were glad to see the end of Puritan rule...

Cambridge in Turmoil

Isaac left the rural peace of Lincolnshire for Trinity College, Cambridge. He arrived to find the university in uproar. It had been the home of Puritanism just as Oxford was of Monarchism. Under Cromwell all royalist heads and fellows had been turfed out. Now they were back in favour, and the round Puritan caps were swiftly replaced by the traditional square mortar board.

1659

1660

THEY HAVE SOLVED AN ANCIENT MATHEMATICAL PROBLEM.

HOW SO?

WHY MILADY, THEY HAVE SQUARED THE CIRCLE.

How did the university receive the 18-year-old who was to be the most illustrious student in its history?

Sizar

Isaac was enrolled as Humphrey Babington's sizar. Babington was the brother of Mrs. Clarke the Apothecary's wife. He was a powerful man, one of the eight fellows who ran Trinity College.

Sizars, poor scholars, were considered the lowest form of university life. They did duty as valets and servants. The other students preferred not to be seen communicating with them except to give orders. In the dining-hall they were only allowed to eat the left-overs.

Luckily Babington was only in residence five weeks a year, so Isaac had plenty of time to improve himself. However, his studies were disturbed by an unruly room-mate. At the age of 18 Isaac was four years older than the average student, who spent a good deal of the day carousing.

Chums

Nicholas Wickins here describes his father John's first days at Trinity and a momentous meeting.

"My Father's first Chamber-fellow being very disagreeable to him, he retired one day into ye walks, where he found Mr Newton solitary and dejected; Upon entering into discourse they found their cause of retirement ye same, & thereupon agreed to shake off their present disorderly Companions & Chum together, which they did as soon a conveniently they could, & so continued as long as my Father staid at College"

The two shared rooms for twenty years. Little is known about Wickins except that he was an invaluable asset to Newton. He not only did the chores but helped with experiments and wrote fair copies. And though Isaac lived at Trinity for twenty-eight years this was his only real friend. Having settled into his rooms it was time to get on with his studies.

Ye Official Curriculume

The official curriculum at Cambridge had been the latest thing when it was laid down by law a century earlier. It began with logic (Aristotelian), ethics (Aristotelian) and rhetoric (Aristotelian) as a foundation for the study of Aristotelian Philosophy and ended with Aristotelian disputations — preferably quoting Aristotle in his own language.

SOUNDS GOOD TO ME

Aristotle of Stagyra, 384-322 B.C.

VERY FAST

QUITE BRISK

But by 1661 Cambridge was no longer advanced. European philosophy had progressed and left it behind. In this intellectual backwater lectureships were not taken too seriously. Few of the university fellows even bothered to tutor.

WET!

ARISTOTLE

PUPIL MONGER!

Common Sense

The world-view held by Aristotle was 2,000 years old — and wrong. The reason that it had outlived the thinking of other ancient Greeks who had come closer to the truth was that it seemed to make more sense.

THE EARTH MOVES AROUND THE SUN.

Aristarchus the Heliocentric, 310-230 B.C.

RUBBISH, THE EARTH IS UNMOVING AND THE CENTRE OF THE UNIVERSE.

Aristotle

Some remnants of history have casually escaped the shipwreck of time. And as flotsam and jetsam the lightest ideas float to the surface while all that is solid and worthwhile sinks.

As nobody was being flung off into space, people thought Aristotle must be right. Especially when he backed up his ideas with a complicated theory of motion. What made things move, he said, was their desire to get to their "proper place"...

Francis Bacon, 1561-1626; Viscount St.Albans and an expert in these matters.

Physics

The word Physics comes from the Greek *physis* which means nature, but not as we understand it. The *nature* of a thing is its end, the reason for which it exists. Those things are *natural* which by continuous movement, originating from an internal principle, arrive at some completion.

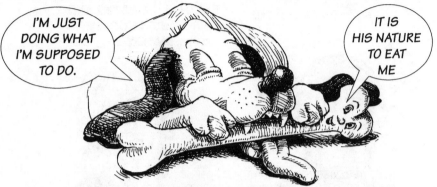

When a dog seizes a bone, the dog moves while the bone remains at rest, and the motion has a purpose, to fulfil the dog's physis or nature. To be eaten by a dog is the nature of the bone.

The universe is divided into two distinct realms. Below, within the so-called sub-lunary sphere, the *natural* motion is a straight line directed towards the centre of the Earth. For example a cannon-ball when fired will shoot off in a straight line *(violent motion)* until it is spent, when it drops vertically seeking its natural place *(natural motion)*.

Crystalline Spheres

In the heavens are the planets, the stars, the Sun and the Moon. Their *natural* motion is circular, perfect, continuous and infinite. Each planet is fixed to a rotating sphere of crystal which revolves around the stationary Earth. Combinations of spheres within spheres rotating in different directions produce the complex paths of the planets across the sky.

*...how build, unbuild contrive
To save appearances, how gird the Sphere
With Centric and Eccentric scribbled o'er,
Cycle and epicycle, Orb in Orb*

IF GOD HAD CONSULTED ME BEFORE EMBARKING UPON THE CREATION, I SHOULD HAVE RECOMMENDED SOMETHING SIMPLER.

Alfonso X, 1221-84

The Athens of Aristotle's day was a democracy, but not for all. Citizens could devote their time to culture because slaves did all the work. This fostered a prejudice. Practical things like experiments were strictly for the lower classes. Rejection of careful observation and measurement separated the ancient 'natural philosophers' from the new 'natural scientists', whose views were in the air at Cambridge when Newton arrived, even if they were not on the curriculum.

Copernicus, who started the new movement, was himself very much of the old school.

Wandering Stars

Claudius Ptolemy, A.D. 120-190

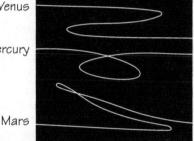

The word planet comes from the Greek meaning 'to wander'. And wander they do, sometimes slowing in their general drift eastward, hanging stationary and even going backwards (retrograde) in strange looping paths.

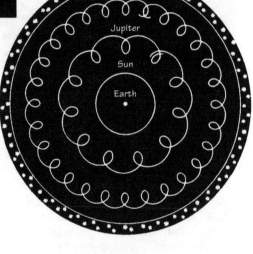

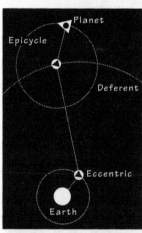

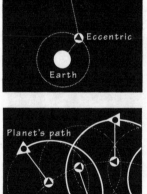

Ptolemy at Alexandria perfected Greek cosmology with a model that enabled the astronomer to 'save the appearances' and with an intricate system of wheels within wheels reproduce the erratic paths of the planets entirely with circles.

Reasonable Arrangement

Nicholas Copernicus was dissatisfied that the planets of Ptolemy's system move with variable speeds.
He agreed with Aristotle that the planets should move with uniform speed in perfect circles.

Nicholas Copernicus, 1473-1543

The Polish astronomer wasn't much of a star-gazer. He didn't even average one observation per year. While at University in Bologna, he heard of the theory of Aristarchus that the Earth moves around the Sun. All of Aristarchus' writings had disappeared, but his ideas survived because Aristotle had taken the time to refute them.

If all the planets circled the Sun instead of the Earth their retrograde motions could be easily explained, but to make this system fit with observations Copernicus still had to make liberal use of Ptolemy's epicycles and deferents.
His new system could hardly be called simplified, needing 46 circles to explain the ballet of the planets compared with 27 for Ptolemy. Nor was it Sun-centred, the centre of the universe being located at the centre of the Earth's orbit, a point in space some distance from the Sun. However, it was to inspire others to find the real answer.

Motions of Mars

IT WOULD BE LIKE POURING CLEAN FRESH WATER INTO A WELL FILLED WITH DIRT, FOR THE DIRT WILL ONLY BE AGITATED AND THE WATER WASTED.

Frightened by the prospect of controversy, Copernicus locked away his book for thirty years. But rumours spread. In 1539, a young German professor of Mathematics and Astronomy named Rheticus presented himself at Copernicus' door in Frauenburg to see for himself.

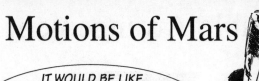

ON THE EXTREME OUTSKIRTS OF THE EARTH.

Georg Joachim Rheticus, 1514-76

Rheticus was allowed to read the manuscript. It took ten weeks to plough through. Mars was especially perplexing. Rheticus prayed for help. An angel appeared, seized him by the hair and alternately banged his head against the ceiling and then the floor, saying: "These are the motions of Mars". Shaken but resolute, Rheticus undertook to publish the book in Nuremberg, but halfway through the vice-squad pounced.

UGH! THE ITALIAN PERVERSION.

COMES FROM STUDYING TOO MUCH GREEK.

On the Revolutions

Publication was taken over by the co-founder of Lutheranism, Osiander, who added his own preface —

"Let no one expect anything certain from astronomy lest he depart from this study a greater fool than he entered it."

Copernicus received the first copy from the printers on his deathbed. The introduction didn't help his health and he departed not only the study of astronomy, but also this world, simultaneously.

Copernicus's fear of publication and controversy had been unnecessary, the first edition never sold out. The theory didn't upset the Pope in the least, but it did offend Luther's common sense.

Martin Luther, 1483-1546

A NEW ASTROLOGER WANTS TO PROVE THAT THE EARTH MOVES AND REVOLVES, INSTEAD OF THE SKY, THE SUN AND THE MOON. JUST AS IF SOMEBODY MOVING IN A CARRIAGE MIGHT HOLD THAT HE WAS SITTING STILL, AT REST, WHILE THE EARTH AND TREES WALKED AND MOVED. RIDICULOUS!

HE HAS JUST FORMULATED AND DISMISSED THE THEORY OF RELATIVITY.

On the Revolutions of the Heavenly Spheres wasn't a revolutionary book; Copernicus didn't try to throw out Aristotle and Ptolemy but to save them. But importantly, he was the first to advance a consistent system with a moving Earth, even if it was wrong.

The Key

One of the very few who actually read Copernicus' book was a German astronomer.

I BELIEVE I'VE FOUND THE KEY TO THE UNIVERSE.

this is it ➤➤

Johannes Kepler, 1571-1630,

Johannes Kepler had been born prematurely, and became a sickly child with short sight and multiple vision. He was plagued with sores, constant stomach and gall bladder problems, piles, rashes, mange and worms. He thought he was a dog.

HIMMEL!

Kepler's interest in the heavens began early. Before he was nine he had seen both a comet and an eclipse of the Moon. He often wondered why there were only six planets, and on July 9th, 1595, he discovered the reason.

Perfect Solids

PYTHAGORAS DISCOVERED THE PUTTING TOGETHER OF THE COSMIC FIGURES

These are the 'perfect solids', whose faces are all identical and regular. The laws of geometry limit their number to five.

Being perfectly symmetrical, a perfect solid can rest within a sphere with each of its corners resting on the inside surface.

Similarly, a perfect solid can contain a sphere which rests against the centre of every face.

Kepler was entranced by the idea that he could insert the five solids between the orbits of the six planets.

It Fits

At the age of 26 Kepler solved the riddle of the Universe and published his findings in the book *Mysterium Cosmographicum*, The Cosmic Mystery.

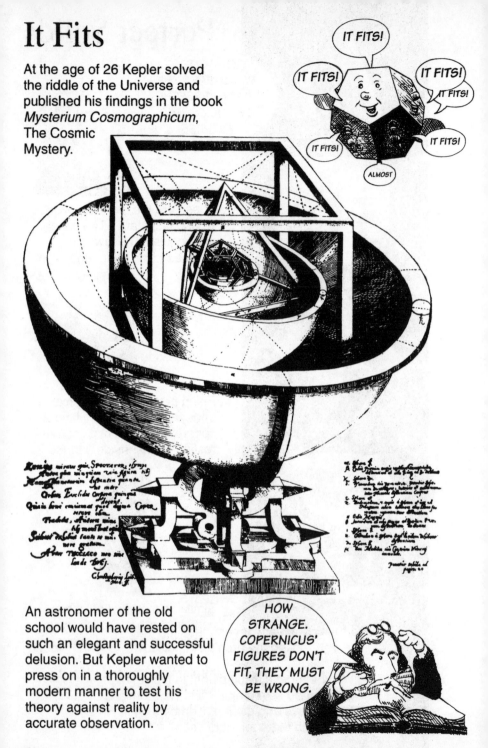

An astronomer of the old school would have rested on such an elegant and successful delusion. But Kepler wanted to press on in a thoroughly modern manner to test his theory against reality by accurate observation.

Meet Tycho

In 1598 Kepler was thrown out of Austria in a purge of Protestants. Without a job, he willingly accepted an invitation to live with the noble Danish astronomer Tycho Brahe at Benatek Castle outside Prague, especially since he needed Tycho's observations to perfect his *Mysterium*.

Mrs. Barbara Kepler

> PRAGUE, WHOSE STENCH WAS ENOUGH TO DRIVE BACK THE TURKS

Jeppe the clairvoyant dwarf.

Tycho had the best observations, and thus the material to build the new edifice. He only lacked the architect to discover the truth hidden deep within them.

Kepler was given the most difficult planet to tackle, Mars, for which Copernicus had required seven epicycles. He promised to have the solution within eight days and even placed bets on it. He was destined to wrestle with Mars for eight years.

Super Nova

As a teenage stargazer Tycho had been appalled to find the standard authority on planetary motion, the Alphonsine Tables, a whole month out, and Copernicus several days wrong. He decided to collect precise, up-to-date, continuous observations.

> I'VE BUILT THE FIRST REALLY ACCURATE INSTRUMENTS AND I'LL WRITE MY OWN TABLES.

> HE IS ABLE TO TAKE OBSERVATIONS THROUGH HIS NOSE, NEEDING NO OTHER INSTRUMENT.

Tycho had lost his nose in a debate with a well-armed mathematician, so he had a new one made up in metal.

Tycho Brahe, 1546-1601

Ursus, his arch rival.

In 1572 Tycho noticed a new star in Cassiopeia, so bright that it could be seen in daylight. He couldn't believe his own eyes.

Closely Observed Stars

All over Europe the leading astronomers brought to bear the most accurate techniques available...

Get a bit
of thread

Hold it at
arm's length

Line up with
nearby stars.

Sit still for
several hours.
See if it moves.

THREAD!
MY QUADRANT IS 38
FEET ACROSS, BUILT OF
BRASS AND OAK.

ONE
OF TYCHO'S
INSTRUMENTS IS
WORTH MORE
THAN MY ENTIRE
FAMILY'S
FORTUNE

SO
WHAT?

Aristotle taught that the fixed stars were perfect and therefore unchanging. Growth and decay could only take place this side of the Moon, within the so called sub-lunary sphere.

Why was Tycho so excited? Because if the new star could be proved to be located amongst the fixed stars, the heavens had changed, and that in contradiction of Aristotelian dogma. The next year he published a book, *De Stella Nove*. Page after page of "hard obstinate facts" established the new star beyond doubt.

Perfect and Unchanging

In 1577 he dealt another blow to Aristotle by proving that the great comet of that year was not a sub-lunar phenomenon, but lay "at least six times" as far away as the Moon.

Tycho's spirit was triumphant but his flesh was weak. His bladder gave way during an epic banquet in 1601. He died in a delirium repeating over and over, the whole night: "Let me not seem to have lived in vain."

THE ONLY CONSTANT IS CHANGE

Kepler immediately appropriated Tycho's vast body of observations. He also took over as Imperial Mathematician to the Holy Roman Emperor Rudolf II, heir to Charlemagne.

Except for the occasional pause to dash off horoscopes for the Prague nobility, Kepler could now forge ahead in developing the three laws of planetary motion that in time would become the basis of Newton's laws.

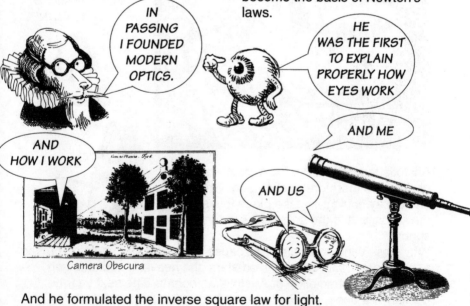

IN PASSING I FOUNDED MODERN OPTICS.

HE WAS THE FIRST TO EXPLAIN PROPERLY HOW EYES WORK

AND ME

AND HOW I WORK

Camera Obscura

AND US

And he formulated the inverse square law for light. But all the while it was Mars that dominated his life...

The Sleepwalker

MARS TOOK SUCH A HOLD OF ME THAT I NEARLY LOST MY MIND.

NEARLY?

The calculations of Mars' orbit alone cover 900 folio pages of tiny handwriting. Kepler got into a terrible tangle, making errors that cancelled each other out, explaining the errors incorrectly and labouring up blind alleys for years before stumbling over the truth in what Arthur Koestler described as "the most amazing sleepwalking performance in the history of science".

WHAT MAKES KEPLER, THE MYSTIC OF THE COSMIC MYSTERY SUCH A STICKLER FOR DETAIL?

HE HAS INTRODUCED REAL PHYSICAL CAUSE AND EFFECT INTO THE ABSTRACT GEOMETRY OF THE SKIES

Reckoning Made Easier

Help was at hand for these astronomical calculations

A SCOTTISH BARON HAS DONE AN EXCELLENT THING BY TRANSFORMING ALL MULTIPLICATION AND DIVISION INTO ADDITION AND SUBTRACTION.

IT IS NOT FITTING FOR A PROFESSOR OF MATHEMATICS TO MANIFEST CHILDISH JOY JUST BECAUSE RECKONING IS MADE EASIER.

Kepler's Maths teacher, Michael Mästlin, 1550-1631

Logarithms did the trick, developed by the Laird of Merchiston during twenty years of isolation in his castle near Edinburgh.
The principle is to pair an arithmetic series (top)
with a geometric series (bottom).

0	1	2	3	4	5	6	7	8
1	2	4	8	16	32	64	128	256

To multiply 8 by 32 convert numbers from the bottom line to the top.
8 becomes 3 and 32 becomes 5. Add 3 to 5, answer 8.
Convert 8 from the top line. The answer to 8 times 32 is 256.

John Napier, Eighth Laird of Merchiston, 1550-1617

Napier himself held that his most important work was the book *A Plaine Discovery of the Whole Revelation of Saint John.*

I PROVE IN EUCLIDEAN FASHION THAT THE POPE IS THE ANTI-CHRIST AND THAT THE WORLD WILL END IN 1786.

Calculating rods known as Napier's bones.

The Shape of the Universe

AH, WHAT A FOOLISH BIRD I HAVE BEEN.

After eight years of calculation Kepler finally came to the conclusion that Mars' path is quite simply not a circle.

Mars' orbit is oval — an ellipse, with the Sun at one focus. At a stroke Kepler reduced the seven Copernican epicycles to one elegant curve.

Kepler likened ridding astronomy of the epicycles to Hercules' cleansing of the Augean stables.
He left behind "only a single cartfull of dung" — the ellipse.

Copernicus had deposed the Earth from its central position and sent it wandering off among the other planets, but Kepler alone challenged Aristotle's belief that the *perfect* planets could only move in perfect circles. Kepler changed the shape of the universe.

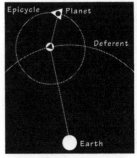

Ptolemy's Universe

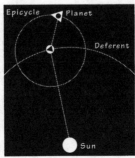

Copernicus' Universe

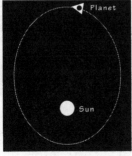

Kepler's Universe

Kepler's Second Law

Kepler threw out Copernicus' other central assumption, that the planets travel with constant speeds.

He noticed that as a planet recedes from the Sun it slows down and as it approaches closer it speeds up. Convinced that there must be some harmonious relationship, Kepler eventually realised that in spite of the speed varying, the area swept out by the planet in a given time was constant.

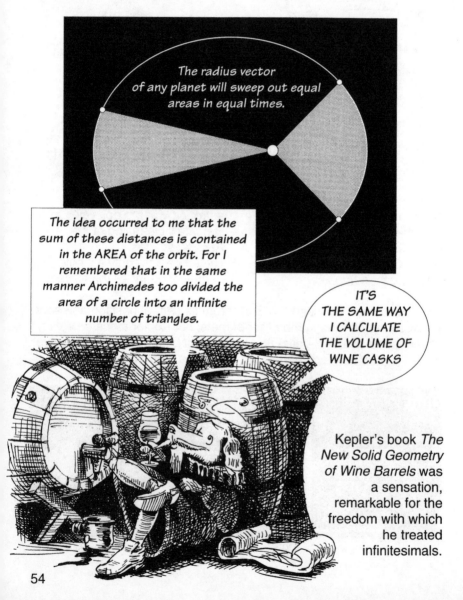

The radius vector of any planet will sweep out equal areas in equal times.

The idea occurred to me that the sum of these distances is contained in the AREA of the orbit. For I remembered that in the same manner Archimedes too divided the area of a circle into an infinite number of triangles.

IT'S THE SAME WAY I CALCULATE THE VOLUME OF WINE CASKS

Kepler's book *The New Solid Geometry of Wine Barrels* was a sensation, remarkable for the freedom with which he treated infinitesimals.

Harmony of the Worlds

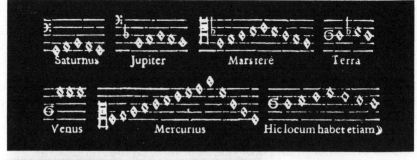

THE HEAVENLY MOTIONS ARE NOTHING BUT A CONTINUOUS SONG FOR SEVERAL VOICES

To notate the "music of the spheres", Kepler did not hesitate to compare any attribute, seeking any and every harmonic relationship between the five planets.

Buried amongst the musical notes was a curious little relationship. It seems that the squares of the periods of revolution (T) of any two planets are as the cubes of their mean distance from the Sun (r).

	Year (T)	T squared	Orbit (r)	r cubed
Mercury	0.2408	0.0580	0.388	0.0584
Venus	0.6152	0.3785	0.724	0.3795
Earth	1.0000	1.0000	1.000	1.0000
Mars	1.881	3.5378	1.524	3.5396
Jupiter	11.862	140.71	5.200	140.61
Saturn	29.457	867.72	9.510	860.09

This was to become — even though he didn't know it himself — Kepler's Third Law. It is the key to the stability of th for it indicates in what way the motions of the five p mathematically interdependent. The book containin universally ignored. Three days after the completior *Harmony of the Worlds*, the Thirty Years War broke

The Rudolphine Tables

After extending his laws for Mars to the other planets and their satellites (Kepler's own word), thereby charting the solar system roughly as we know it today, Kepler had one last task — to present Tycho's observations in table form.

LET ME NOT SEEM TO HAVE LIVED IN VAIN.

Named in honour of Emperor Rudolf II, work on the *Rudolphine Tables* dragged on. Now, in Linz, he tried to compete his life's work as Catholics and Protestants fought on the roof of his print shop.

He had to take time off to turn defence lawyer and save his 73-year-old mother from being burnt as a witch.

JUST PROMISE NOT TO TURN THE JUDGE INTO A TOAD

THEY HAD ME IN CHAINS FOR 14 MONTHS, SON

Kepler Passes Away

THIS BOOK WILL LAST FOR CENTURIES.

The printers caught fire and the book was destroyed, but Kepler managed to escape with the manuscript. He moved to Ulm for another try. *The Rudolphine Tables* were finally completed in time for the Frankfurt Book Fair of 1627. Kepler set up a stall and sold them himself.

In 1629 he invented science-fiction. *Somnium* is the tale of a journey to the moon, strictly following the laws of physics. He describes the intense acceleration needed to take off from Earth, the equal pulls of Earth and Moon in midflight ... weightlessness.

HE UNDERSTANDS THE GRAVITY OF THE SITUATION

Kepler's salary was now 11,817 florins in arrears. So he set off on horseback (and on foot on account of piles) to try to collect money owed him. After weeks on the road he fell ill, and three days later died pointing his finger alternately at his head and at the sky.

Skybound was the mind, earthbound the body rests.

The Wrangler

Meanwhile, in Italy, a maths teacher from Pisa, Galileo Galilei, was determined to get above his station.

He had been a tiresome obstinate student, earning himself the nickname "Wrangler". He left university without a degree, more because of his abrasive personality than his scholarship. He even had to leave Tuscany altogether after a clash with a member of the ruling Medici family.

This is the lamp which inspired Galileo in 1581, it was installed in 1587.

At the age of 17 he came upon the principle of the pendulum clock during a service in Pisa cathedral.

Galileo Galilei, 1564-1642

While still in his twenties he had realized that a body ten times as heavy doesn't fall ten times as fast, as Aristotle had taught.

The Leaning Tower of Pisa from which Galileo never dropped a cannon-ball.

IN 1586, I PUBLISHED PROOF THAT HEAVY AND LIGHT BODIES FALL AT THE SAME SPEED.

Jan de Groot

Father of the Telesco[pe]

Galileo was so impressed by the new discoveries being made in science that he decided to make them his own.

> I MAKE IT MINE AND CHASE AWAY THE OWNER.

He made quite a fortune by manufacturing and selling a military compass.

It had been invented fifty years before in Germany.

Next he set his sights on the telescope.

> I INVENTED THE TELESCOPE IN 1600

> I INVENTED IT IN 1589

> OH YEAH! WELL I SOLD IT TO THE DOGE OF VENICE FOR 3,000 SCUDI IN 1609

Johannes Lilippershey (ND)

Giovanni Battista della Porta (I)

Galileo, foster-father of the telescope

When confronted with the evidence that he had not invented the telescope, Galileo replied that any simpleton might well have discovered the telescope by accident but only he, Galileo, could have discovered it by means of reasoning, for that required much greater ingenuity, therein lying the real merit of the invention.

Starry Messenger

seen through this recent invention the telescope, was
...ed. That was yet another blow to the Aristotelian notion of
the "crystalline perfection" of the heavens. After twenty years of
teaching Aristotelian cosmology Galileo started to contradict it.

THE MOON IS NOT ROBED IN A SMOOTH AND POLISHED SURFACE

Galileo's own drawing.

THE MOON IS IN FACT ROUGH AND UNEVEN, COVERED EVERYWHERE, JUST LIKE THE EARTH'S SURFACE, WITH HUGE PROMINENCES, DEEP VALLEYS AND CHASMS

BUT I'VE ALREADY SAID THAT

Thos. Harriot (GB)

In 1610 Galileo published his observations in book form. Written in
Latin, it was elegant, witty, sensational and short. The first edition of
the *Starry Messenger* was sold out in a matter of days. Overnight,
Galileo's name became a household word. He even became the
darling of intellectual churchmen in Rome. That coming man,
Cardinal Maffeo Barberini, showered him with praises.

First and Alone

Galileo stunned the public (and other astronomers) by claiming a series of startling discoveries.

The four satellites of Jupiter

First observed by Simon Mayr.

Venus' phases

Predicted by Castelli.

The triple form of Saturn

Spiral Galaxy in Andromeda

First observed by von Greenhausen.

Sun-spots First observed by Christoph Scheiner (SJ).

First Warning

Increasingly taken with his own importance, Galileo was making some very important enemies amongst the Jesuits. He then proceeded to try to push Copernicus' heliocentric system down the Church's throat.

IT CANNOT BE HELPED THAT IT WAS GRANTED TO ME ALONE TO DISCOVER ALL THE NEW PHENOMENA IN THE SKY, AND NOTHING TO ANYONE ELSE.

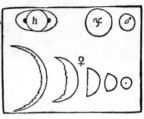

COPERNICUS IS ALL VERY WELL IN THEORY...

... BUT GALILEO WANTS TO REWRITE THE BIBLE TO FIT HIS OWN IDEAS.

Because he could not prove that the Earth moved around the Sun, Galileo challenged the Church to prove otherwise!

The inquisitor Cardinal Bellarmine, who had burnt Giordano Bruno at the stake for heresy in 1606, was put on the case.

To speak hypothetically as Copernicus spoke is excellent good sense and runs no risk whatsoever ... If there were real proof we should have to reinterpret the Holy Scriptures, but none has been shown to me.

Cardinal Robert Bellarmine, d. 1621

Galileo was officially forbidden to hold or teach Copernicus' System as true unless he could prove it.

Two Grand Systemes

In 1623 Galileo's admirer Barberini became Pope Urban VIII. There was a fresh liberal atmosphere abroad in Rome. Galileo had a series of six audiences with the new Pope and was encouraged to write about Copernicus as long as he confined himself to theory. The Pope even came up with the title.

GALILEO, YOU MAY KISS MY TOE.

The book, in colloquial Italian, is couched in the popular form of a dialogue, just like his father's *Dialogues on Ancient and Modern Music.*

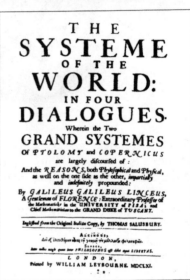

THE SYSTEME OF THE WORLD: IN FOUR DIALOGUES. Wherein the Two GRAND SYSTEMES Of PTOLOMY and COPERNICUS are largely difcourfed of : And the REASONS, both Phylofophical and Phyfical, as well on the one fide as the other, impartially and indefinitely propounded: By GALILEUS GALILÆUS LINCEUS, A Gentleman of FLORENCE : Extraordinary Profeffor of the Mathematicks in the UNIVERSITY of PISA and Chief Mathematician to the GRAND DUKE of TUSCANY. Englifhed from the Original Italian Copy, by THOMAS SALUSBURY. LONDON, Printed by WILLIAM LEYBOURNE. MDCLXI.

There are three characters.
Salviati, a brilliant savant, who delivers Galileo's theories.
Sagredo, a reasonable layman who quite sensibly is convinced by Salviati's arguments.
Simplicio, a simple idiot who defends Aristotle; he is invariably proved wrong. The conversations take place during the course of four days.

The **First Day** is taken up with refutations of the Aristotelian view of the cosmos and the incorruptibility of the heavens.
The **Second Day**, is dedicated to overcoming objections to the Earth's motion.
The diagram on the right shows that no body, regardless of its weight, could be flung off the surface of the Earth by its rotation, no matter the speed.

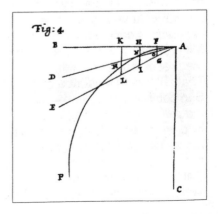

Fig: 4

Going Round in Circles

The **Third Day** describes the Copernican system's superiority to Ptolemy's.
However, the system Galileo describes so admiringly is not that of Copernicus, but a gross travesty. He seems never to have read *On the Revolutions of the Heavenly Spheres*.

> PTOLEMY IS DISEASED AND COPERNICUS HAS THE CURE.

Ptolemy

Copernicus

Galileo is convinced, like Copernicus and Aristotle, that the only way to move is in circles. Natural motion is circular because it always arrives back where it started and therefore can continue. Straight motion is by nature infinite. It is impossible that any thing should have by nature the principle of moving in a straight line, or in other words, towards a place where it is impossible to arrive.

> IT'S JUST THAT THE CIRCLES ARE SO BIG THAT WE ONLY SEE A SHORT PIECE WHICH LOOKS STRAIGHT.

The planets stay in their perfectly circular orbits because things can only move circularly. Anything that does not move circularly is necessarily immovable, nothing but rest and circular motion being suitable for the preservation of order.
That any weird influence coming from the Sun could affect the paths of the planets (as Kepler believed) is superstitious nonsense. Because they do not follow circular paths, comets are banished. Galileo pronounced them optical illusions caused by atmospheric effects. He called them Tycho's *Monkey Planets*.

Conclusive Proof

The **Fourth Day** proclaims that the tides are produced by the rotation of the earth. Kepler's correct explanation, published seven years before, that the tides are caused by the Moon's influence, is ridiculed.

> THE THOUGHTS OF KEPLER TEND RATHER TO THE DIMINUTION OF THE DOCTRINE OF COPERNICUS THAN TO ITS ESTABLISHMENT

I foretell his coming to a bad end.

grrr!

> I'm surprised at Kepler. He has lent his ear and given his approval to the Moon's dominion over the waters, to occult properties, and such puerilities.

> THE TIDES ARE CAUSED BY THE EARTH'S MOVEMENTS — AND AS THERE ARE TIDES, THEREFORE THE EARTH MOVES.

Galileo is totally wrong about the tides, and his logic is false. But this is his ultimate weapon, conclusive proof that the Earth moves. The Pope had tried to point out the weakness of his argument, saying it was neither true nor conclusive.

> How can I prove the Earth moves when those who must be convinced show themselves incapable of following even the simplest and easiest of arguments?

The First Family

The book started off on the wrong foot. The frontispiece engraved by Stefan Della Bella shows three figures discussing the merits of their systems.

On the left is the old master Aristotle, in the middle his disciple Ptolemy carrying a model of nested geocentric spheres, and on the right is Copernicus holding a symbol of his own heliocentric universe. But the sting in the tail is on the ground between their feet. There lies a coat-of-arms with three fish biting one another's tails.

This was immediately recognizable as a parody of the three bees in the Barberini family crest, and a sly dig at the nepotism of the Pope, who was well on the way to turning the Vatican into a family business, filling key posts with his relatives.

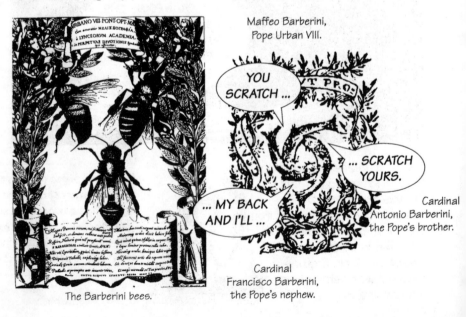

The Barberini bees.

Maffeo Barberini, Pope Urban VIII.

Cardinal Francisco Barberini, the Pope's nephew.

Cardinal Antonio Barberini, the Pope's brother.

Beyond Understanding

Galileo had been given permission to write on promising an impartial treatment of both sides of the argument. However, when the *Dialogue* appeared it was quite specific about those who didn't agree with Galileo.

THEY ARE MENTAL PYGMIES. HARDLY DESERVING TO BE CALLED HUMAN.

But if the book starts with a snigger it ends with a bang. The fourth day winds up with a doctrine from a certain *most eminent and learned person before whom one must fall silent*. Simplicio addresses Galileo's *alter ego* Salviati ...

THAT'S IMPARTIAL?

"*I confess that your hypothesis on the flux and reflux of the tides is far more ingenious than any of those I have ever heard; still, I esteem it neither true nor conclusive, but I know that if you were asked whether God in his infinite power and wisdom might confer upon the element of water the reciprocal motion in any other way, you would answer that he could, and in many ways, some beyond the reach of the intellect.*"

Even if the movement of the Earth explained the tides it did not follow that the tides proved the movement of the Earth. God might well have produced tides by quite other means beyond our understanding. This was the very argument that the Pope had advanced during the audiences that preceded the book. Galileo put it into the mouth of Simplicio, the man who was never right.

The first copies of *Dialogue* arrived in Rome in August 1632. Within a matter of hours the Pope discovered that his friendship had been betrayed.

GALILEO HAS DECEIVED **ME!**

Now Urban VIII was possibly the most vain megalomaniac in Italy besides Galileo himself. This was the man who melted down the bronze ceilings of the Pantheon for cannon. Of whom it was said "What the barbarians didn't do, Barberini did". This was personal insult. He wasn't about to take it lying down.

A Trial

Galileo appeared before the Inquisition on April 12, 1633.

His attempted defence further shows the contempt he has for his opponents' intelligence. A compulsive debater, he tries to convince the Inquisition that the book says the reverse of what it does. He maintains that the book proves the opposite of the Copernican opinion by showing the arguments of Copernicus to be weak and inconclusive.

HAS HE GONE MAD?

After a life of getting away with everything, it gradually dawns on Galileo that the mental pygmies were not to be swayed by his brilliant rhetoric. His feelings of invincibility desert him. He panics and offers to completely rewrite the book. He publicly denies ever having believed Copernicus. Here was certainly not the stuff of martyrs.

I HAVE NEVER MAINTAINED OR DEFENDED THE OPINION THAT THE EARTH MOVES

IS THIS SUPPOSED TO BE A JOKE?

He was treated with elaborate courtesy and respect during the trial, and housed in a luxurious five-room suite with servants. But it was inevitable that he be found guilty and the *Dialogue* proscribed. Galileo was sentenced to prison and to repeat the seven penitential psalms once a week. He was however allowed to get his daughters to recite the psalms for him (both had become nuns), and he was allowed to go home.

Two New Sciences

After twenty years spent tied up with fierce polemics, Galileo was finally muzzled and confined to his home. At last he was free to finish the book he had always promised.

Dialogues on Two New Sciences was published in 1638. **Salviati**, **Sagredo** and **Simplicio** make a four-day return performance.

The **First Day** treats the resistance which bodies offer to fracture.

The **Second Day** concerns the causes of cohesion.

LOOKS FAMILIAR.

Oresme

The **Third Day** is dedicated to uniform motion and naturally accelerated motion.

The **Fourth Day** describes violent motion and projectiles.

Bonaventura Cavalieri submitted an idea to Galileo for his opinion. It was a mathematical proof that the curve followed by a projectile would be one of the conic sections, a parabola. Galileo's opinion was that Cavalieri ought to wait for a more opportune moment before publishing it.

In fact Galileo liked the idea so much that the opportune time and place turned out to be the **Fourth Day** of his own dialogues.

The Finger

The Church should tell us how to go to Heaven, not how the heavens go.

Galileo's right index finger is on display at the Museo di Storia della Scienza, Piazza dei Guidici 1, Florence.
Open: Mon., Wed. & Fri., 1400-1700 h.

If Galileo never knew the inside of a prison cell, wasn't tortured and didn't defy authority, if he didn't drop cannon-balls off the Leaning Tower of Pisa and didn't invent the telescope or even the thermometer, what on earth did he do?

Francis Bacon compares the pursuit of science to a military campaign where force is not enough but must be backed up with artfulness. Galileo's policy of "murder your brothers and steal their valuables" may be criminal but was necessary to build the empire of science. If all the various discoveries he collected had remained dispersed with their owners they would not have helped the general progress of experimental science.

Raffaele Caverni in *History of the Experimental Method*, 1891, provides several well-mixed metaphors on the subject.
"By ruthlessly pruning the tree of science Galileo concentrated the nutritive sap from all the subterranean roots into one bud, himself."

But Galileo's true legacy may just be his pupils, bullied, exploited, but faithful ... Cavalieri, Torricelli, Castelli, Aggiunti, Viviani, Borreli, Paolo and Candido de Buono. They created a myth out of a man.

Like Clockwork

NATURE IS AN AUTOMATON.

René Descartes (F), 1596-1650

Descartes, like Galileo, believed that the book of nature was written in mathematical language. But he went the whole hog and said that nature was just a machine. Cartesian physics is often referred to as Mechanistic, meaning that it uses no other principle for explanation than concepts dealt with in mechanics (the science of the behaviour of matter under the action of force). Descartes attached his own meaning, namely, that which can be imitated in a mechanical model.

Matter can only affect matter by colliding with it.

Animals having no souls have no feelings.

Including plants.

All living things are just intricate machines.

Vivisection gets carte blanche.

A doctor is just a mechanic.

OUCH!

DESCARTES OBVIOUSLY HAS NEVER HAD DEALINGS WITH SHEEP. TRICKY BEAST THE SHEEP.

Principles of Philosophy

> All things which we very clearly and obviously conceive are true (only observing, however, that there is some difficulty in rigidly determining the objects which we distinctly conceive).

Descartes

There is no sense in asking a machine what its purpose is, it having no other than that which its maker gives it.

When Newton arrived in Cambridge, "he found such a stir about Descartes, some railing against him and forbidding the reading of him as if he had impugned the very gospel. And there was a general inclination of the brisk part of the University to use him."

Matter can only affect matter by colliding with it, so, armed with exact laws of motion and impact, we should be able to predict and explain everything that happens in nature.

Descartes book *Principia Philosophiae* (1644, The Principles of Philosophy) was a triumph of fantastic imagination which happens, unfortunately, never once to have hit on a correct explanation.
But Descartes did construct a completely new philosophical edifice from the ground up, something no one had attempted since Aristotle.

Newton swallowed him whole, for here was someone who promised escape from Aristotle. Aristotle was simply irrelevant, Descartes was something to get your teeth into.

The Whys and Wherefores

There are different ways of looking at motion...

WHY DOES A BODY MOVE?

SEEKING ITS RIGHTFUL PLACE.

Aristotle

HOW DOES A BODY MOVE?

$$S = \frac{(V_1 + V_2)T}{2}$$

Galileo

WHY DO THINGS STOP?

THAT'S NEW

Descartes

THE TENDENCY TO MAINTAIN MOTION IS AN ENDEAVOUR TO CONTINUE UNIFORM MOTION IN A STRAIGHT LINE

The Vortex

Descartes' Second Law of Nature states: "A body has a tendency to persist in a state of rest or uniform motion in a straight line".
It follows that the natural path of a planet is a straight line, not the circle of Galileo. A planet would fly off at a tangent unless there was some other influence forcing it to deviate from its natural path. It is the pressure of the vortex that holds a planet in its curved orbit.

IT MAKES NO SENSE TO SAY THAT CIRCULAR MOTION IS **SUPERIOR** TO MOTION IN A STRAIGHT LINE.

Here is the path of a comet buffeted by its passage through the vortex.

Descartes imagined a Universe entirely built of matter and motion. The original large chunks of matter created by God were left to rub against each other, producing fragments of three different sizes. Splinters of the second (celestial) matter are centrifugally forced into conglomerates (the Sun and stars) by huge vortices of finest primary particles of subtle matter (aether). The coarser tertiary dross forms the Earth and planets.
The centrifugal outwards urge of the spinning celestial matter is felt reflected back as Gravity. The pressure of the whirling celestial matter is observed as the light emitted by the Sun and stars.

Perpetual Motion

Newton was intrigued, if not entirely convinced, by Descartes' explanations. He began to systematically question them. If light is caused by pressure, "we should be able to see in the night as well as or better than ye day", he concluded.

THE VORTEX IS PRESSED WITH DIFFICULTIES.

A MAN GOEING OR RUNNING WOULD SEE IN Y^e NIGHT.

If Descartes' vortex really exists, and gravity is caused by the descent of subtle matter ...

...THEN Y^e RAYS OF GRAVITY MAY BE STOPPED BY REFLECTING OR REFRACTING THEM AND PERPETUAL MOTION MAY BE MADE.

Whereas Descartes thought of space as completely filled with matter, the *Plenum*, Newton agreed with Gassendi, Lucretius and Democritus that space is empty, the *Void* in which atoms move.

The Book of Nature

In the painting *The School of Athens* by Raphael, Plato indicates the sky and ideal forms while Aristotle gestures at the ground to emphasize the importance of learning from nature. Aristotle had been seen as the more go-ahead of the two, as he was forty-five years Plato's junior and had been his pupil. In the Middle Ages Aristotle's physics was blessed by Thomas Aquinas (1225-74) and became the dogma of the Church. A reaction set in. Aristotle, who for the previous generations represented realism and close contact with nature, became perceived to be an old-fashioned philosopher who had lost touch.

IT'S QUANTITY THAT COUNTS.

NO, QUALITY IS BEST.

Plato, 429-347 B.C.

Aristotle, 384-322 B.C.

Galileo put the Neo-Platonist creed like this, "Philosophy is written in that great book, the universe, which stands continually open to our gaze. But the book cannot be understood unless one first learns the language and can decipher the letters of which it is composed. It is written in the language of mathematics, and its characters are triangles, circles, and other geometric figures without which it is humanly impossible to understand a single word of it; without these, one strays as in a darkened maze."

Newton was drawn into the Cambridge Neo-Platonist circle. They were determined to replace Aristotle's stifling influence with that of the more mathematical Plato.
Leading light of the group was Isaac Barrow, the Lucasian Professor of Mathematics.

"Get your eyes to help your ears! Make experiment the companion of reason!"
– Barrow

A Universal Philosophy

The Neo-Platonists felt that Descartes had gone too far in eliminating the spiritual and immaterial from the physical world, leaving only matter and motion. Unlike Descartes, Alchemists got their ideas by questioning nature directly — by experimentation.

Isaac Barrow 1630-77

"Descartes has inverted the order of philosophising, it seemed good to him not to learn from things, but to impose his own laws on things. He first collected truths which he thought suitable. Principles he had framed without consulting nature."

"Alchemy is the only Art which might be able to complete and bring to light not only medicine but also a universal Philosophy."

From Barrow, Newton learned that Alchemical philosophy is the equal of mathematics, and Alchemical experimentation equals any anatomy or botany. From another member of the group, Henry More, he learned respect for antique and esoteric literature and acquired the belief that the real secrets lay hidden there.

BARROW IS THE BEST SCHOLAR IN ENGLAND.

Charles II

77

Spirit of Nature

Henry More was born in Grantham and had been Dr. Clarke the Apothecary's tutor. Newton had already met More's philosophical position among Clarke's "great parcel of books".

DESCARTES JUST WON'T SEE SENSE.

Mere impact of one particle on another cannot account for all the phenomena of nature, but a directing Spirit of Nature, the great quarter-master-general of Providence, must needs intervene.

Henry More 1614–87

More originally welcomed the Cartesian philosophy as "the most sober and faithful in the Christian world". But, as time went by, his suspicions grew that Descartes's complete divorce of matter from spirit would end badly. More engaged in fruitless correspondence with Descartes to get him to qualify his position. His criticisms of Cartesian *Mekanik* were contained in the book *Immortality of the Soul*, 1659. As time went by the implications of the mechanical theories began to emerge. The Neo-Platonists' worst fears were incarnated in the person of Thomas Hobbes.

Nasty, Brutish and Short

The bête noire of the Neo-Platonists was that arch-materialist Thomas Hobbes, who was engaged in bitter mathematical debate with Barrow.

I HAVE SO SQUARED THE CIRCLE.

HOMO...

Thomas Hobbes, 1588-1679

Hobbes thought of men as utterly devoid of any sense of right or wrong. Lacking the capacity to help themselves, they were as much automata as cog-wheels in a clock. As men were incapable of making moral decisions they had to be made for them. He proclaimed his philosophy in the book *Leviathan*.

...HOMINE...

"In the state of nature there is no property, no justice or injustice, only war. And such a war as is of every man against every man. Force and fraud in war are the two cardinal virtues. Men are forced to co-operate by reasons of pure selfishness. Law and morality are no more than organized violence. The dominant urge of man is self preservation, manifested primarily as fear, continual fear and danger of violent death; and the life of man solitary, poor, nasty, brutish and short."

...LUPUS.

Descartes' theories clearly held dangerous consequences. But for the moment they offered the only way to break Aristotle's stranglehold.

Blazing Star

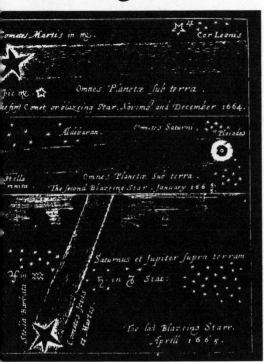

In November 1664 a comet appeared, of a faint, dull, languid colour, and its motion very heavy, solemn and slow, foretelling a judgement, slow, but severe, terrible and frightful.

It seemed that the comet passed directly over the city of London and that so very near the houses that it was plain it imported something peculiar to that city alone.

At the beginning of December two men, said to be Frenchmen, died of the plague at the upper end of Drury Lane. The bubonic plague previously known as *The Black Death* had returned. Before the epidemic had passed a fifth of London's population would be dead.

Swift & Furious

Hard on its heels came another comet, this one flaming, swift and furious. It promised yet another hotter punishment for the town.

The commission called by Parliament to investigate the cause of the Plague knew where to lay the blame. God was obviously pretty upset by the publication of the works of Hobbes.

Isaac's own remedy for the plague...

THOMAS HOBBES'S BOOKS ARE FORBIDDEN FORTHWITH.

Lucatello's Balsam

Indications: The measles, plague and smallpox.

Contents: Turpentine, the best damask rosewater, beeswax, olive oil and sack, flavoured with a pinch of red sandalwood and a dash of oil of St. John's wort.

Instructions: Mix a quarter of an ounce in broth, take it warm and sweat afterwards. For the bite of a mad dog apply also to the wound.

Business came to a standstill. The court left town. Those who could fled to the countryside. In 1665 the plague arrived at Cambridge. As a result the University closed and was to remain so for two years. Newton returned home to Woolsthorpe, liberated from the constraints of the Aristotelian curriculum.

A Trifling Book

With the university closed, Isaac was given his head to roam the field of learning wherever his fancy took him. First he plunged into Mathematics.

Plato advised starting with mathematics because it is a science which proceeds very cautiously and admits nothing as established until it is rigorously demonstrated.

LIKE A HIGH SPIRITED YOUNG HORSE WHO MUST FIRST BE BROKE IN PLOWED GROUNDS & THE ROUGHEST AND STEEPEST WAYS OR COULD OTHERWISE BE KEPT WITHIN NO BOUNDS.

Isaac's interest had originally been aroused at Stourbridge Fair.

He buys a book on astrology, but is unable to cast one of the figures in it. So he buys Euclid's Geometry, looks up the theorems he needs in the index — and finds them obvious. His appetite for geometry whetted, he begins devouring Schooten, Oughtred, Wallis and Descartes.

The Quantity of Quality

Line of intensity

Latitude (intension)

Area = Quantity of quality

Longitude (extension)

Uniform motion

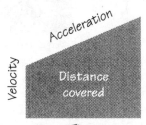

Velocity

Acceleration

Distance covered

Time

Uniformly difform motion

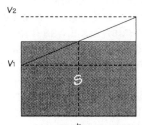

v_2

v_1

S

t

$$S = \frac{(v_1 + v_2)t}{2}$$

Nicholas Oresme, 1323-82: Bishop of Lisieux.

Hipparchus (161-126 B.C.) had been the first to define points on a map by means of coordinates (latitude and longitude).
Nicholas Oresme applied this graphic tool to the medieval problem of the intension and remission of forms. This method of treating increase and decrease of intensity was called *Calculationes*.
It was applied to variables like temperature, light, weight and even love. But the quality which best benefited from this treatment was motion.

By regarding instantaneous velocity as latitude and time as longitude Oresme obtained the diagram for uniformly difform motion. The distance covered is given by the area produced.
An example of motion that is regularly changing is the acceleration of a falling body. Oresme proposed that the distance covered during a uniformly changing motion in a given time is equal to that covered by a uniform motion during the same time at an average of the speeds.

I'M GOING TO INVENT THAT IN 1632

Galileo

Fatiguing the Imagination

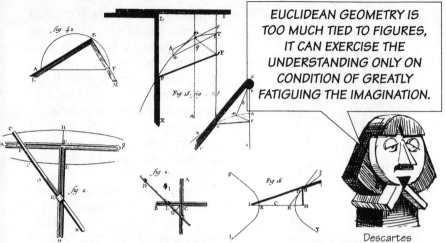

EUCLIDEAN GEOMETRY IS TOO MUCH TIED TO FIGURES, IT CAN EXERCISE THE UNDERSTANDING ONLY ON CONDITION OF GREATLY FATIGUING THE IMAGINATION.

Descartes

In **Geometry**, curves were well known from the Greeks but only by their methods of construction. There was no uniform way of manipulating them and there was no general way of expressing the characteristics of a curved line. Meanwhile **Algebra** was, according to Descartes, "full of confusion and obscurity, calculated to embarrass".

DON'T FORGET, THE WORLD WILL END IN 1786

Descartes combined the best of both into **analytic geometry**. Not only could any equation be represented geometrically but classes of curve when written in equation form were seen to correspond to classes of equation. This alliance of geometry and algebra had been used earlier, but it was from Descartes that the new system became widely known.

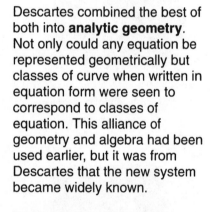

Napier had used a curve to show the relationship between logarithms and numbers. This curve was a Hyperbola, but it proved impossible to square by exhaustion.
In order to calculate the irrational numbers (like the base of logarithms) it was necessary to expand them as infinite series which could then be summed term by term until the desired degree of accuracy was reached.

THE ETERNAL SILENCE OF THESE INFINITE SPACES TERRIFIES ME.

Pascal had pointed out that the coefficients for an expansion could be obtained from an array still known as Pascal's triangle, in which each number is the sum of the nearest two numbers in the row above.

```
            1
          1   2   1
        1   3   3   1
      1   4   6   4   1
    1   5  10  10   5   1
  1   6  15  20  15   6   1
1   7  21  35  35  21   7   1
```

Blaise Pascal, 1623-62: inventor of the wristwatch, syringe, pocket calculator and omnibus.

Several useful expansions were already known. Newton was familiar with John Wallis' infinite series for approximating π, found in his book *Arithmetica Infinitorum*.

WALLIS' BOOK IS A SCAB OF SYMBOLS.

Thos. Hobbes

$$\frac{\pi}{2} = \frac{2}{1} \cdot \frac{2}{3} \cdot \frac{4}{3} \cdot \frac{4}{5} \cdot \frac{6}{5} \cdot \frac{6}{7} \text{ etc.}$$

Following Wallis, Newton began to fill in the spaces between the numbers in Pascal's triangle. At last he devised a general method for finding coefficients of infinite series without the triangle even for negative or fractional terms. Now called the *Binomial Theorem*, Newton's formula looked like this :–

EXTRACTION OF ROOTS IS MUCH SHORTENED BY THIS THEOREM.

$$(P+PQ)^{m/n} = P^{m/n} + \frac{m}{n} AQ + \frac{m-n}{2n} BQ + \frac{m-2n}{3n} CQ + \frac{m-3n}{4n} DQ + \text{etc.}$$

Hyperbole

THIS OPERATION MAY BE CONTINUED AT PLEASURE, YE FARTHER THE BETTER.

Isaac attacked the Hyperbola,

$$y = \frac{1}{(1+x)}$$

and with the infinite series thus produced he started to calculate logarithms to fifty-five decimal places.

I am ashamed to tell you to how many places of figures I carried these computations having no other business at the time.

Infinite series were no longer only approximating devices but fully equivalent to finite functions. The *Binomial Theorem* legitimized the use of the infinite. The *horror of the infinite*, which had haunted mathematicians since Zeno, was banished and calculus unleashed.

WHATEVER COMMON ALGEBRA PERFORMS BY EQUATIONS OF FINITE TERMS MAY ALWAYS BE PERFORMED BY INFINITE SERIES.

Within a year of first opening Euclid, Isaac has absorbed the entire body of existing mathematics. From now on he is on his own.

A State of Flux

Archimedes, Oresme, Kepler, Galileo and Descartes, all had calculated areas by adding up infinitesimal parts. Newton, considering areas to be generated by continuous motion of points, lines and planes, took the momentary increase in area at the point in question and from that rate of change he calculated the area.

Finding areas under curves had been thoroughly explored by the ancients. But the problems taxing modern mathematicians — lenses, curved mirrors, atmospheric distortion, orbits of planets and the motion of the Moon to locate ships at sea — are tangent problems.

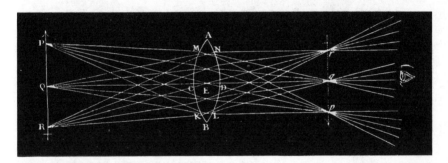

Newton realized that tangent and area problems are identical but the inverse of each other. So he created a fusion. A general theory of equations, of tangents, of infinite series and areas. A universal solution for all curves. Newton had invented the *calculus*.
He called it **Fluxions**.
The name comes from the idea of *flow*. The variable mathematical quantity generated by motion is a **fluent** and its rate of change a **fluxion**.

"Calculus is the art of numbering and measuring exactly a thing whose existence cannot be conceived."

Voltaire

The Fall

If the Earth is rotating why does the apple fall down and not sideways or upwards as Aristotle predicts it should?

thud!

THAT REMINDS ME, I MUST FEED THE CAT.

A body in circular motion strives constantly to recede from the centre, like a stone on a string as it is whirled about. Only the pull of the string keeps it from flying off at a tangent.

To calculate the force needed to hold a body in a circular path, Newton calculates the force needed to return a body to its original position by bouncing off the sides of a square. This same relation holds as the number of sides is progressively doubled.

So if a body were reflected by the sides of an equilateral circumscribed polygon with an infinite number of sides (i.e. a circle) the force of the reflections is to the force of the body's motion as all those sides (i.e. the circumference) to the radius.

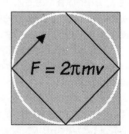

$$F = 2\pi mv$$

Using a pendulum Newton worked out that the force exerted by the rotation of the Earth to throw us off into space was only 1/350th of the force of gravity sticking us onto the surface.

Pretty Nearly

It occurs to him that the power of gravity (which causes the apple to drop from the tree to the ground) is not limited to a certain distance from the ground but that this power must extend much higher than is usually thought.

> WHY NOT AS HIGH AS THE MOON, AND IF SO THAT MUST INFLUENCE HER MOTION AND PERHAPS RETAIN HER IN HER ORBIT.

He asks himself what the endeavour of the Moon to recede from the centre of the Earth would be. He calculates it to be 1/4,000th the force of gravity at the surface of the Earth.

And when he substitutes Kepler's third law into his new formula for rotational force he finds that...

> THE ENDEAVOURS OF RECEDING FROM THE SUN WILL BE RECIPROCALLY AS THE SQUARES OF THE DISTANCES FROM THE SUN.

"Just like my inverse square law for light."

The value for the Moon of 1/4,000th was suggestive but it was only approximately the value of 1/3,600th he would have expected using 60 Earth radii as the distance from the Earth to the Moon. As Newton puts it, it agrees "pretty nearly" but that's not good enough so he puts the idea aside for another day.

"And this was the sole mortal who could grapple, Since Adam with a fall, or with an apple."

Lord Byron

The Lucasian Chair

Returning to Cambridge after the plague, Isaac rises swiftly through the ranks. In 1667 Newton is elected as a fellow of Trinity with an income of £13 6s 8d, he is one of the favoured few who can stay on. In 1669, Isaac Barrow resigns the Lucasian professorship in Newton's favour. Now Newton has the highest post in the college, aside from the Master himself, and £100 a year.

The duties consist of one lecture per week. His future is secure.

"As long as I can avoid lèse-majesté, heresy, schism, homicide, notable theft, adultery, fornication, perjury or intolerable negligence."

But he is largely ignored. Isaac Barrow reports that when the young professor lectured, "so few went to hear him and fewer still understood him, that ofttimes he did in a manner, for want of hearers, read to the walls."

A Bare Bodkin

In 1665 Robert Hooke published his *Micrographia*. It contained excellent drawings of fleas by Christopher Wren, but it was not just a book about the microscope. It advanced a complete theory of light based on the mechanical philosophy of Descartes. Isaac was curious to test Hooke's conclusions.

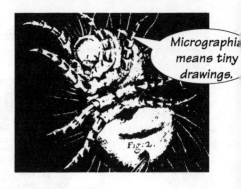

Micrographia means tiny drawings.

Fig: 2.

"I push a bodkin betwixt my eye and ye bone as neare to the backside of my eye as I can and pressing my eye with ye end of it there appear several white dark and coloured circles, which circles are plainest when I continue to rub my eye with the point of ye bodkin."

Isaac does not spare himself in researching the phenomenon of colour. Practising what the Neo-Platonists preached, his practical experimentation is carried as far as staring at the Sun until he almost goes blind and sticking blunt needles in behind his eyeball to see the effect.

He finds that Hooke, and indeed everyone else from Aristotle to Descartes, has misunderstood the fundamental nature of light.

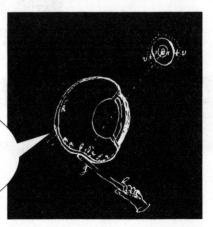

TO RECOVER I SHUT MYSELF UP IN MY CHAMBER MADE DARK FOR THREE DAYS TOGETHER. I BEGAN IN THREE OR FOUR DAYS TO HAVE SOME USE OF MY EYES AGAIN.

Darkness & Light

The current view was that colour was created by the mixture of light and darkness. Red was pure white light with a little admixture of darkness. The last step before complete extinction of light by darkness was Blue.

CONSIDER A BOOK. THE MIXTURE OF BLACK TEXT WITH THE WHITE PAGE WOULD LOOK COLOURED, NOT GREY, AS IT GETS FURTHER AWAY..??

The experimental evidence...

Descartes passed a beam of light through a gall prism and examined the result on a bit of paper two inches away.

Hooke passed a beam of light through a glass beaker full of water and projected it on a piece of paper two feet away.

Newton projected a beam of light through a prism at a wall twenty-two feet away!

According to Descartes there should be two spots, one Red and the other Blue. Instead Newton sees a continuous spectrum. A rainbow of colours eight inches long.

Paradoxicall Assertion

COLOURS ARE SIMPLE, WHITE IS A MIXTURE.

Colours are not produced by disturbing pure white light. For Newton finds it is the colours which are pure. They are seen not by modifying white light but by splitting it into its components

I PERSWADE MYSELFE THAT THIS ASSERTION ABOVE THE REST APPEARS PARADOXICALL, AND IS WITH THE MOST DIFFICULTY ADMITTED.

According to Newton's findings the spectrum of colours is divorced from the grey scale of dark to light. When coloured bodies reflect light onto a piece of white paper, the paper takes on that colour. If he is correct, paper ought to appear paradoxically white, not black, in the light reflected from a black shiny body.

AND IT DOES!

Flaws in the Glass

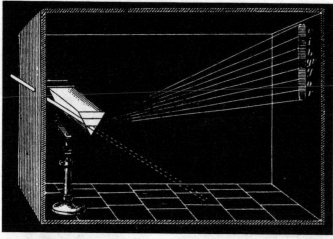

violet
indigo
blue
green
yellow
orange
red

To make completely sure, he would need to prove that the spectrum wasn't just created by any flaws in the glass. So he bought other prisms at the Stourbridge Fair of 1668.

He placed a second prism upside down behind the first. In this way changes produced by any eventual irregularities in the glass should be doubled whereas effects produced by the triangular shape of the prism should be cancelled out.

The resulting image was a pure white round spot of light, exactly as if it hadn't passed through any prism at all. White light could be re-created by mixing all the colours. The spectrum wasn't merely a chance effect produced by the character of the glass, but was to do with the nature of light itself.

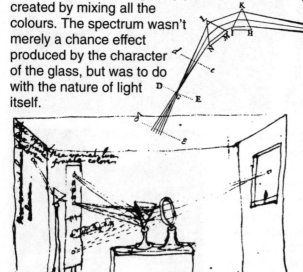

After many years of experimenting in order to remove any shadow of doubt, Newton refines his method into what he calls, echoing Francis Bacon, the *Experimentum Crucis*.

The Crux of the Matter

Fixing boards with small holes, he could isolate a ray of a single colour from the spectrum. This ray of a single colour was unchanged by passing through the second prism. A blue ray remained blue. A red remained red, but was bent at less of an angle.

Intrigued by the colours in soap bubbles and other thin films Newton developed an experiment to measure them. By pressing a glass lens of known curvature onto a sheet of glass he caused a series of coloured circles, now known as *Newton's Rings*, to appear.

"Bodies", said Newton, "are composed of transparent particles the thickness of which determines the colours they reflect, just as the thickness of the film of air between the lens and the glass determines the colour of the rings."

Isaac's measurements, made by eye, are accurate to a hundredth of an inch.

On Reflection

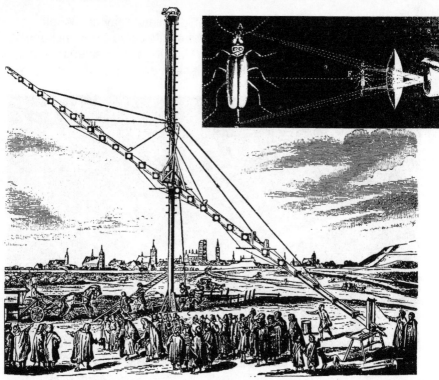

Telescopes in the 1600s were reaching up to two hundred feet long but the coloured fringes produced by refraction through their lenses were limiting their effectiveness. The lenses had spherical surfaces, so Newton tried grinding lenses with various other curves to get rid of the unwanted colours. Trials convinced him that it was impossible, so he abandoned the *glassworks*. He did realize that **reflection** would be free from the aberrant colours as the light bounced off the material and did not pass through it.

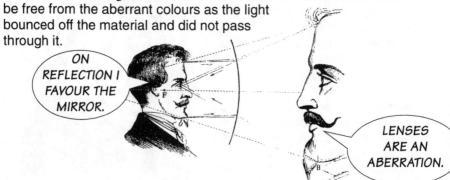

ON REFLECTION I FAVOUR THE MIRROR.

LENSES ARE AN ABERRATION.

He set about building a telescope around a mirror instead of a lens.

Telescope

The resulting reflecting telescope was only six inches long but magnified by forty diameters, which was more than a conventional refracting telescope six foot long would do.

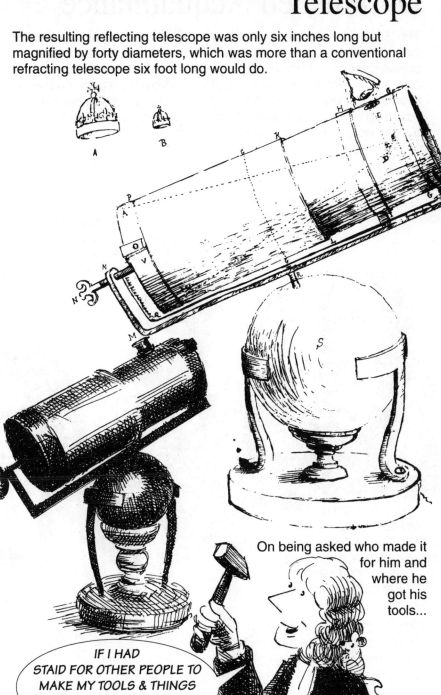

On being asked who made it for him and where he got his tools...

IF I HAD STAID FOR OTHER PEOPLE TO MAKE MY TOOLS & THINGS FOR ME, I HAD NEVER MADE ANYTHING OF IT.

Diminished Acquaintance

The 'Sixties are the most creative years of Newton's life. He is on his own in mathematics and optics and is on the brink of his supreme creation, the law of universal gravitation. The fact that all this has passed unnoticed, even when proclaimed from the lecture podium, seems not to trouble him. Newton desires rather nothing but the peace and quiet to pursue his studies. But in the 'Seventies the outside world begins to intrude...

Newton grudgingly agrees for Collins to publish his formula for calculating annuities, but only anonymously.

Going Public

Newton was not the first to face this dilemma...

Pythagoras

Archimedes

Copernicus

Einstein

...nor would he be the last.

Agonizing over whether to *go public* with his discoveries, or not, would torture him on and off for the rest of his years.

Promotion of Knowledge

The Royal Society for the Promotion of Natural Knowledge, founded in London in 1660, strongly supported the New Science. In 1671 it heard of Newton's reflecting telescope and asked to see it. The King himself was impressed by the usefulness of the device.

"Dashed clever, eh, what?"

"I propose we elect Mr Newton."

Sam Pepys

Seth Ward

I SPY A HEAVENLY BODY... M'LADY PORTSMOUTH.

King Charles II

WE MUST SECURE THIS INVENTION FROM THE USURPATION OF FOREIGNERS.

Oldenburg, Secretary of the Royal Society.

In January 1672, Newton was elected a fellow of the Royal Society.

Odd Detection

In appreciation of the honour shown him, Newton offers the Royal Society what lay behind the telescope, his theory of light and colours. It is well received except by Robert Hooke, Curator of Experiments.

NEWTON HAS JUST FINISHED SOME THINGS I STARTED.

Hooke claims to have made a little tube an inch long that magnifies better than a fifty-foot telescope. He being too busy with the aftermath of the Great Fire to develop it. As to Newton's theory of colours, the main part of it is in his *Micrographia*.

Robert Hooke, 1635-1703

Newton, who describes his theory of colours as "the oddest if not the most considerable detection which hath hitherto been made in the operations of nature", explodes with anger.

Hooke was one of the key figures in the rebuilding of London after the Great Fire. He was "crooked and of low stature" and as he got older more and more deformed, with an eye "full and popping".

HOOKE'S THEORY IS NOT ONLY INSUFFICIENT BUT UNINTELLIGIBLE.

It was the first of what would be many bitter disputes involving Newton over the next half-century. He could not stand criticism even when it was justified, and when it wasn't...

Well, what would you expect with a childhood like mine?

Slave to Philosophy

But Hooke wasn't the only problem. Letters began to pour in from all over Europe with objections to Newton's theory of colours.

MR. NEWTON ATH NOT SHOWN HE NATURE AND DIFFERENCE OF COLOURS.

Christiaan Huygens (ND) 1629-95

MR NEWTON'S EXPERIMENTS ARE WRONG, AS WE CAN SEE WITHOUT EVEN NEEDING TO TRY THEM.

Linus, Lucas and Gascoines, the English Jesuits in Liège,.who would continue to harass Newton for over ten years.

He was driven to distraction by frustration and rage. With his nerves in shreds he admitted...

IN HUNTING FOR A SHADOW, I HAVE SACRIFICED MY PEACE OF MIND. A MATTER OF REAL SUBSTANCE.

It takes four years of infuriating argument before the Royal Society gets round to repeating the experiments to confirm his conclusions.
Newton swings between outbursts of towering rage and total silence, often declaring that he wants nothing further to do with the "promotion of philosophy".

For I see a man must either resolve to put out nothing new or become a slave to defend it.

Put Out

In a fury, Newton pleads with the secretary of the society, "I desire that you will procure that I may be put out from being any longer fellow of the Royal Society."

"I hope you will not find it ill if you find me ever refusing doing any thing more in that kind ... and prevent any objections or letters that may concern me."

The only subject on which Isaac will deign to correspond with Oldenburg is Cider!

WHICH LIQUOR I WISH, WITH YOU, TO BE PROPAGATED FAR AND NEAR IN ENGLAND.

After years of feuding, Newton writes a major book on optics to establish his theory and settle the disputes once and for all. It is all but finished when he goes out for a walk one March morning in 1678, leaving a candle burning on his table.

Henry Oldenburg, 1615-77

THAT'S A LOAD OFF MY MIND! I'LL JUST POP OUT FOR A BREATH OF FRESH AIR.

Utterly Lost

"When Mr. Newton returned from chapel and had seen what was done, everyone thought that he would have run mad, he was so much troubled thereat that he was not himself for a month after." — Abraham de la Pryme.

The book "had the ill luck to perish and be utterly lost". It is the last straw. Isaac completely abandons optics.

But he has other things on his mind. Not content with mathematics, physics, astronomy and philosophy, Isaac has been wading into alchemy and theology — revealing not only the forces of nature but also, in secret, the history of the church.

WHY IN SECRET?

BECAUSE MY BELIEFS ARE VEHEMENTLY HERETICAL!

IF I'D KNOWN OF THEM, HE'D HAVE BEEN OUT ON HIS EAR.

Barrow, now Master of Trinity College

Subtle Spirits

His experiments with light have caused him to turn more and more away from Descartes' description of physical reality, the standard mechanical system of nature. He has never been happy with its separation of mind and body, its elimination of the spirit. It conjured up a boring world, free of sound, smell, colour or feeling.

Newton is reported as being, "intent on Chimicall Studies and practices, and beginning to think mathematical speculations to grow at least nice and dry, if not somewhat barren."

Hermetic philosophy because of its experimental nature had a better basis than Descartes' theories. Subtle spirits might just modify Cartesianism in the right direction.

Sir Isaac carried his enquiry very far downwards into the ultimate components of matter, as well as upwards towards the boundless regions of space.

Newton needed to explain the behaviour of the tiniest bodies in order to complete his universal system. Dealing in substances, essences, souls and virtues rather than matter in motion, Isaac, the Alchemist, was searching for no less than the structure of the universe.

Beyond Human Art

He very rarely went to bed till two or three of the clock, sometimes not until five or six, sleeping about four or five hours. ...

What his aim might be I am not able to penetrate into, but his pains, his diligence at these times makes me think he aims at something beyond the reach of human art and industry.

They who search after the philosopher's stone are obliged to a strict and religious life

Humphry Newton, no relation

Isaac plunges in with typical thoroughness and experimental fervour. He studies ancient alchemy as nobody before (or since). His Alchemical papers finally totalled over a million words (and they're still being examined today!). He tastes a great variety of heavy metals and other toxic substances in his experiments.

"Hmm, delicious."

I wouldn't be surprised if that's not what unbalanced his mind.

Prisca Sapientia

What's that funny smell?

His laboratory was in the garden, connected to his room by a staircase. Especially in Spring and at Fall of Leaf the furnaces would burn for weeks on end, scarcely going out, neither night nor day.

He would sometimes, tho' very seldom, look into an old and mouldy book

Newton had a profound belief in *prisca sapientia*, an original wisdom given to the Ancients. He thought that in the earliest times God had imparted the secrets of natural philosophy and true religion to a select few. The knowledge was subsequently lost but traces could still be found hidden in myths where it would remain unnoticed by the vulgar. Many mysteries had been deliberately disguised to guard them from minds not fit to receive them.

Newton turned to the most esoteric of Alchemical books where he believed the real secrets to be hidden.

NATURE'S ACTIONS ARE EITHER VEGETABLE OR PURELY MECHANICAL.

The Earth a Vegetable

From Alchemy Isaac grasped the principle that Nature is not a mechanism but a living being. All things decay and all things are reborn. "Nature is a perfect circulatory worker." This is Isaac's way of looking at things ...

This Earth resembles a great animal or rather an inanimate vegetable, it draws in ethereal breath for its daily refreshment and vital ferment and transpires again with gross exhalations.

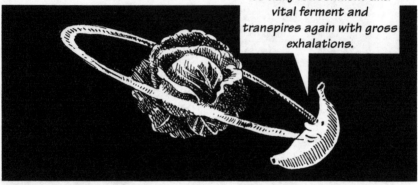

Alchemy leads him on to consider certain puzzling phenomena...

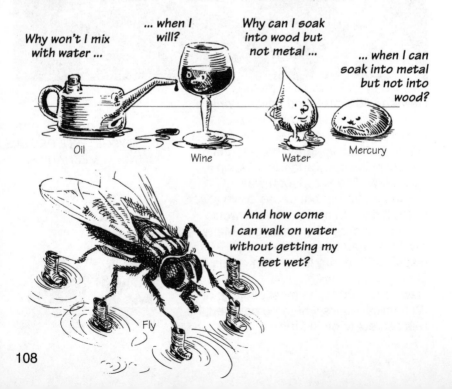

Why won't I mix with water ...

... when I will?

Oil

Wine

Why can I soak into wood but not metal ...

... when I can soak into metal but not into wood?

Water

Mercury

And how come I can walk on water without getting my feet wet?

Fly

Unsociableness

Why do certain substances *socialize* with one another while others *snub* one another? This was something mechanical philosophy could not explain.

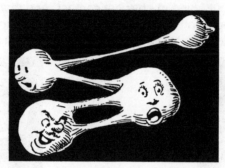

Descartes, like Galileo, had thought the attractions of magnetism and Gravitation *too magical* to consider seriously — part of the mystical twilight world from which the New Scientists had broken free. According to the mechanical crowd, bits of matter could only move and influence one another through direct contact. And the medium through which they communicated was the Aether.

Aether was supposed to be the fifth and most perfect element, and to occupy all space and all bodies. It sounded all right...

Action at a Distance

...but by a series of experiments, with the bob of an eleven-foot pendulum variously filled with sand, metal or wood, Isaac satisfies himself that the aether, a necessity for the mechanical philosophers, does not exist.

The aether has no substance.

Isaac is now sure that particles can and do indeed attract and repel one another without being in physical contact. And that these attractions and repulsions are propagated through empty space without any physical intermediary agent. This is action at a distance.

The mechanical philosophers worked to rid the world of occult forces, attractions and repulsions; the alchemist saw in them the image of the operations of nature.

Letter

In 1679 Isaac was called away. He hurried home to nurse his sick mother. She died in June and he buried her in white wool. Afterwards, he stayed six months in Lincolnshire on family business. Late in the year, out of the blue, a letter interrupted his grief.

OH NO! IT'S HOOKE WANTING ME TO DO HIS SUMS FOR HIM!

It was from Hooke, now, with Oldenburg's death, become secretary of the Royal Society. Hooke's ambition was to achieve what Galileo had failed to do, prove the motion of the Earth. His letter mentioned the orbit of a planet compounded "of a direct motion by the tangent and an attractive motion towards the central body". He wondered what kind of curve such an orbit would be.

Newton wrote back claiming that his affection for mathematics was worn out ...

"In so much that I have long grutched the time spent in that study unless it be perhaps at idle hours sometimes for a diversion."

And although Hooke continued to press him, Newton feigned lack of interest. Assuming that he had shaken the good doctor off, he could return to really important studies, like "Where is God?".

True Worship

Never a man to do things by halves, Isaac spends years absorbing and systematizing the whole of recorded religious history, quite as rigorously and rationally as he pursued his scientific researches.
He comes to the conclusion the Christianity is only a recent offshoot of original true religion, and a corrupt one at that.

THE BIBLE IS JUST ONE SOURCE AMONGST MANY.

Isaac identifies the first *true religion* as that of the Vestal Cult. It worshipped the *God of Nature* in a temple designed to represent the solar system.

"In a temple which is as it were a reflection of God."

The ancients knew the design of the universe by revelation and incorporated it in the floor plan of the temple. Geocentric astronomy was not only wrong, it was blasphemous.

The temple of Vesta, circular and built round a perpetually burning central fire, surrounded by seven lamps, representing the Sun and the seven planets.

Bible a Fake

He spent years studying the plan of the temple of Solomon in Jerusalem, older than any other. He believed that it was a blueprint of paradise.

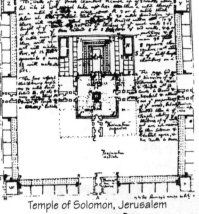

Temple of Solomon, Jerusalem

Isaac is convinced that *true worship* ended when the Egyptians started a trend by creating false gods from their ancestors.

Moses the Lawgiver,
the greatest prophet in history.

He learns Hebrew and retranslates the Bible from original texts.
He discovers that in the 4th century, during a bloody power struggle inside the Christian church, key passages in the Bible were doctored by Athanasius. The falsified text elevates Christ on a level with God and the Holy Spirit in the doctrine of the Holy Trinity. Newton finds that Christ is simply another prophet like Moses, and the worshipping of Christ as God's equal is idolatry.

So Isaac wasn't a Christian at all. His rejection of the Trinity has to be kept secret because the doctrine is central not only to Roman Catholics but also Anglicans.

False Infernal Religion

Newton's abhorrence of the church must have made life with Wickins unbearable. John was a devout man looking forward to the priesthood. Isaac could scarcely share chambers with someone who worshipped "ye Beast & his image".

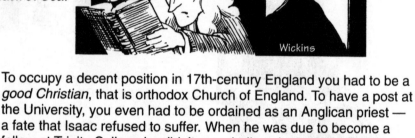

If any shall worship ye Beast and receive his mark, he shall drink the wine of the wrath of God.

YOU GO ON WITHOUT ME.

I'M OFF TO ORDINATION CLASSES ISAAC, YOU COMING?

Wickins

To occupy a decent position in 17th-century England you had to be a *good Christian*, that is orthodox Church of England. To have a post at the University, you even had to be ordained as an Anglican priest — a fate that Isaac refused to suffer. When he was due to become a fellow at Trinity College he didn't even believe in the Holy Trinity! As ordination approached he began to plan his resignation.

I don't think he's cut out to be a priest.

I quite understand, neither am I.

Barrow

King Charles II

In the nick of time he got a Royal dispensation exempting him from taking holy orders. It was arranged by Barrow, the Royal chaplain. But Isaac was obliged to keep his true beliefs secret throughout his life. After twenty years of sharing a room he broke with Wickins.

The Wager

We all suspect an inverse square law governs celestial motion, but no one can prove it.

I can, but it's a secret.

Wren & Halley

Halley, Wren and Hooke wager forty shillings on it and Halley visits Newton to ask his opinion.

What would be the curve described by a planet if the force of attraction towards the Sun were reciprocal to the square of its distance from it?

How do you know?

An Ellipse to be sure.

Why, I have calculated it.

May I see the proof?

It must be here somewhere.

Newton sends Halley a nine-page treatise setting out his calculations. Halley suggests publication, but Newton, as sensitive as ever about appearing in print, replies:

Now I am on the subject I would gladly know the bottom of it before I publish my papers.

And so, in the autumn of 1684, Newton launched into the work that would establish him as the greatest scientist in the history of the world.

115

A Fishy Story

The epoch-making book —
three volumes of obscurely
written Latin — was nearly
never printed at all! The Royal
Society's coffers were empty, all
available money having been
drained by production of an
extravagant edition of
Willughby's *De Historia
Piscium.*

The History of Fishes.

"Yummy, I'll take three copies."

The Royal Society will gladly publish your Principia.

I'll have to pay for it myself.

Edmond Halley, 1656-1742.
Eldest son of a soap-boiler. First to map
the stars of the Southern hemisphere.

Meanwhile, in a London coffee-
house, Hooke is still levelling
accusations of plagiarism against
Newton.

*"He has stolen my idea,
the greatest discovery in
nature that ever was since
the Creation."*

Oooo!

Corr!

Mr. Aubrey

Mr. Lodwick

Dr. Hooke

Little Smatterers

Newton heard of Hooke's secret society and retorted ...

I BELIEVE DR. HOOKE CANNOT PERFORM THAT WHICH HE PRETENDS TO. LET HIM GIVE DEMONSTRATIONS OF IT. I KNOW HE HATH NOT GEOMETRY ENOUGH TO DO IT.

INDEED I COULD IF I SO WISHED. BUT I WON'T.

Hooke never did provide his proof. Nevertheless Newton feared being plunged into fresh exchanges that would embroil him in time-consuming and emotionally upsetting controversy, like the theory of colours did.

He had originally written the third part of the *Principia* in a popular style so that his crucial conclusions would be relatively easy to follow. But when Hooke's attacks resumed, he first threatened to stop publication, then thought better of it and completely rewrote Book III in such a way that it could be easily understood only by those "who had first made themselves masters of the principles established in the preceding Books".

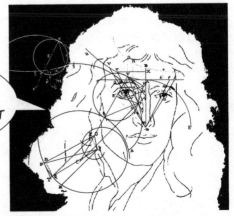

TO AVOID BEING BAITED BY LITTLE SMATTERERS IN MATHEMATICS I SHALL NOT STINT IN COMPLEXITY.

This was clearly a rebuke to Hooke, and Newton rubbed it in by erasing any reference to his antagonist from the *Principia*.

The Best Bits

Copernicus. Keep heliocentricity. Throw out circular orbits and epicycles.

Kepler. Keep the Three Laws, Tides, Gravitation. Throw away his idea of the Sun sweeping the planets round like a broom.

Galileo. Keep behaviour of falling and projected bodies. Throw out circular inertia, circular orbits, the tides.

Descartes. Keep rectilinear inertia. Throw out the Vortex, tides, the Plenum.

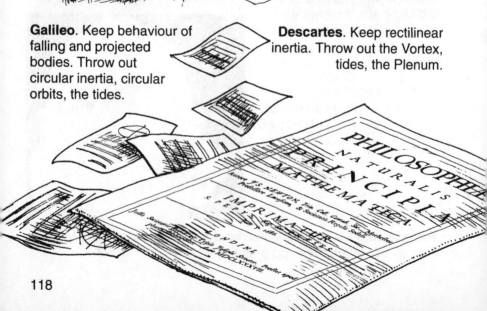

PHILOSOPHIÆ

NATURALIS

PRINCIPIA

MATHEMATICA.

Autore *J S. NEWTON*, *Trin. Coll. Cantab. Soc.* Matheseos
Professore *Lucasiano*, & Societatis Regalis Sodali.

IMPRIMATUR·
S. P E P Y S, *Reg. Soc.* P R Æ S E S.
Julii 5. 1686.

L O N D I N I,

Jussu *Societatis Regiæ* ac Typis *Josephi Streater*. Prostat apud
plures Bibliopolas. *Anno* MDCLXXXVII.

Definitively

The *Principia* starts off with an unshakable foundation on which the mighty edifice of three books will rest. Mass, force and motion are defined, just as Euclid prefaces his Elements with definitions of points, lines and areas.

The whole is constructed in Euclidean fashion with a rigorous logical structure of **Definitions**, **Axioms** (laws), **Propositions**, **Lemmas** (assumptions), **Corollaries** and **Scholia** (explanatory notes).

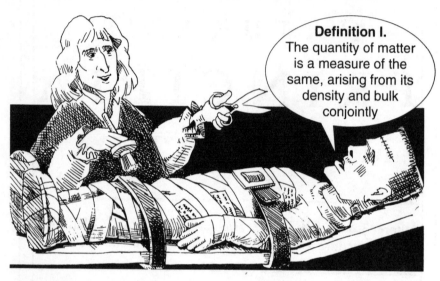

Definition I.
The quantity of matter is a measure of the same, arising from its density and bulk conjointly

Definition V.
A centripetal force is that by which bodies are drawn or impelled, or in any way may tend, towards a point as to a centre.

As opposed to centrifugal force which is the force the body experiences away from the centre.

Centripetal force, that's me.

Centrifugal force is what I feel.

A projectile would fly off in a straight line, if it were not for the force of gravity pulling it to Earth.

If a cannon-ball is fired from a mountain top, it falls into the sea some way distant. The greater the velocity with which it is projected the farther it will go, until, at last, a speed is reached which will carry it right around the Earth, returning to its point of departure from behind. It has gone into orbit. Any faster and it will fly off into space.

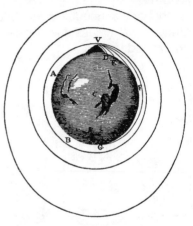

Scholium IV.

ABSOLUTE AND RELATIVE MOTION ARE DISTINGUISHED BY THE EXAMPLE OF A PAIL OF WATER

When the pail is spun around at the end of a twisted rope, the surface of the water rises up the sides. Although the water is at rest relative to the pail, you can tell by the curved surface that it is rotating relative to absolute space.

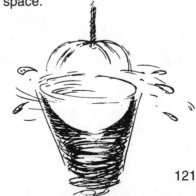

Laws

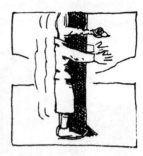

Axiom I.
Every body continues in its state of rest, or of uniform motion in a straight line, unless it is compelled to change that state by forces impressed upon it.

Axiom II.
The change in motion is proportional to the motive force impressed.

Axiom III.
To every action there is always opposed an equal reaction: or the mutual actions of two bodies upon each other are always equal and directed to contrary parts.

# of Motion

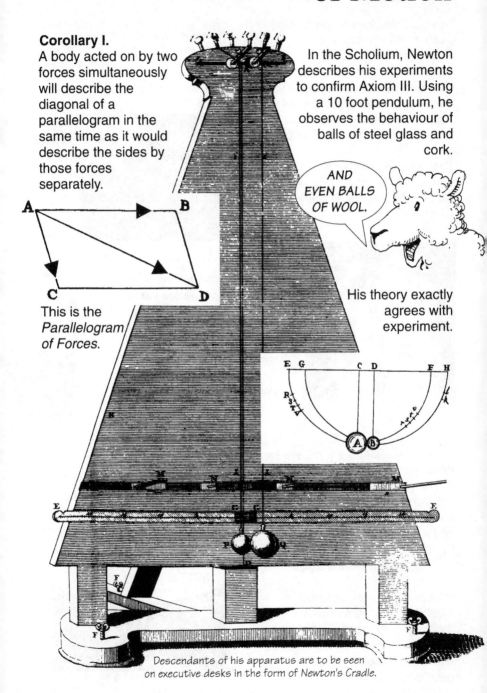

Corollary I.
A body acted on by two forces simultaneously will describe the diagonal of a parallelogram in the same time as it would describe the sides by those forces separately.

This is the *Parallelogram of Forces.*

In the Scholium, Newton describes his experiments to confirm Axiom III. Using a 10 foot pendulum, he observes the behaviour of balls of steel glass and cork.

AND EVEN BALLS OF WOOL.

His theory exactly agrees with experiment.

Descendants of his apparatus are to be seen on executive desks in the form of *Newton's Cradle.*

First & Last Ratios

Now the mathematical method used in the *Principia* is established by a series of Lemmas. To avoid all controversy Newton chooses to demonstrate his propositions with a firmer geometrical underpinning than his *Fluxions*.

The rejection of Cartesian maths was another step on the way to eliminating all traces of Descartes.

Lemma I. Quantities and the ratios of quantities which in any finite time converge continually to equality, and before the end of that time approach nearer to each other than by any given difference, become ultimately equal.

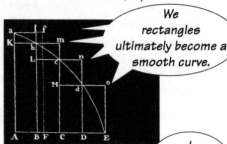

We rectangles ultimately become a smooth curve.

I DON'T FEEL WELL.

"The Ancients' method is more elegant by far than the Cartesian one. His algebraic calculus when transposed to words would prove so tedious and entangled as to provoke nausea, nor might it be understood. But they accomplished it with simple proportions."

Newton sidesteps the infinite by talking not of indivisibles, but *evanescent* divisible quantities*; not the sums and ratios of divisible parts but always the *limits* of sums and ratios, not the ultimate ratio but the *limit* towards which it converges.

However this new method of *First and Last Ratios* was thoroughly modern, Euclid would have been shocked at what had become of his geometry.

Curves are traced by points in motion in short *moments* of time. As the *moment* becomes shorter, Newton takes the value...

...NOT BEFORE IT VANISHES, NOR AFTER, BUT THAT WITH WHICH IT VANISHES.

pop!

*evanescent = about to vanish

124

A Proposition

Proposition I.

Newton shows that motion under the influence of a central force is subject to the law of areas.

This he first demonstrates for motion in a straight line. A line between a moving body (following the path *A,B,C*) and a fixed point (*S*) sweeps out equal areas in equal times.

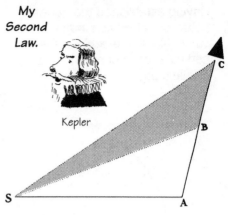

My *Second Law.*

Kepler

Then he deflects the straight motion by a series of blows. Using the parallelogram of forces, he finds that the law of areas is still obeyed.

$$ABS = BCS = CDS \text{ \&c.}$$

If the number of triangles is increased and their breadth decreased ad infinitum, the ultimate path of the motion becomes a curve

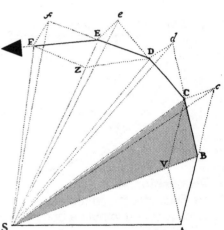

If a body is pulled continuously towards a point, its path will be bent into a curve and its motion will obey Kepler's law of areas.

Proposition VI.

If any curvilinear figure be given, and therein a point *S* is also given, to which a centripetal force is continually directed, that law of centripetal force may be found, by which *P* will be continually drawn back from a rectilinear course, and being detained on the perimeter of that figure, will describe the same by a continual revolution.

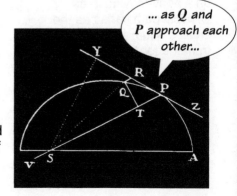

... as *Q* and *P* approach each other...

125

In Eccentric Conics

Having established the general behaviour of motion in any curve, Newton examines the particular case of motion in an elliptical path.

Elliptical paths, that's my First Law.

Kepler

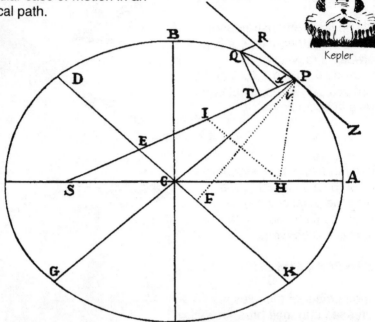

Proposition XI.
He finds that the centripetal force required to hold a body on an elliptical path is as the square of the distance *SP*. That means that the attractive force on a body moving in an elliptical path varies as the inverse square of its distance from the focus of the ellipse.

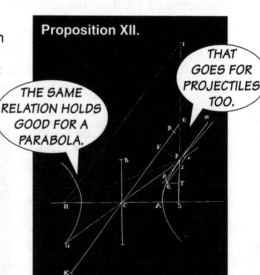

Proposition XII.

THE SAME RELATION HOLDS GOOD FOR A PARABOLA.

THAT GOES FOR PROJECTILES TOO.

Pound Shock Horror Error

The One Pound Note issued by the Bank of England in 1978 bears a portrait of Isaac Newton on the reverse. Newton sits under the apple tree at Woolsthorpe with his reflecting telescope and a prism placed on the garden table beside him. On his lap the *Principia* lies open at Proposition XI. But in the diagram that dominates the design the Sun is in the wrong place!

The note was withdrawn from circulation in 1984.

To:-
Chief Cashier,
Bank of England

Svartensgatan
116 20 Stockholm,
Sweden.
April 19 1983.

Dear Sir,

It can hardly have escaped your, or someone else's, notice that the design on the reverse of the one pound note is flawed.

I refer to the placing of the "Sun" at the centre (C) of the diagram purporting to represent Proposition XI from Newton's "Principia": this in direct contradiction of Kepler's First Law (pub. 1618), which states that planetary orbits are elliptical, with the Sun at one of the foci (diag. S). Newton could not possibly have derived his Inverse Square Law of Gravitation from this diagram which you have rightly identified as being fundamental to his work.

In consideration of the misleading effect this mistake could have upon school children and others, and as a reflection on the Bank of England's reputation, have you considered withdrawing the note to correct the error, which is not a little

Spherical Bodies

"It might be that the proportion that was accurate enough at greater distances would be wide of the truth near the surface of the planet, where the distances of the particles are unequal, and their situations dissimilar."

For the case of bodies long distant from one another, their various parts can be considered equidistant, and the effects of force on their various particles parallel.

The situation of a body only a few feet above the Earth's surface is clearly quite different. The apple will be attracted downwards but also sideways.

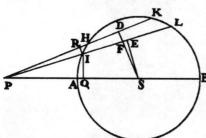

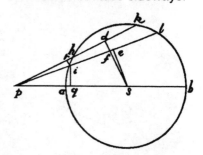

Proposition LXX.

Newton considered the individual forces arising from the infinitely numerous particles of two hollow spheres. He proves that the forces between the two spheres increase or decrease in proportion to the distance between their centres. It is as if the whole mass of the sphere were concentrated in a point at its centre.

So it follows that the previous propositions, about motions in conics around an attracting point, are therefore equally true when an attracting sphere (like a planet) is placed at the focus.

And this is a noteworthy fact.

Resisting Mediums

With pendulums, Newton examines Descartes' Plenum, "an aetherial medium extremely rare and subtle, which freely pervades all bodies". It ought to produce some measurable effects. Newton failed to find any, the resistance is "nil or wholly insensible".

Remembering to take into account the resistance of the thread which is certainly very considerable.

The experiments are made with pendulums in air, under water, in mercury and even in hot oil.

He drops glass globes filled with air or mercury 220 feet from the top of St. Paul's cathedral.

The results are shattering for Descartes.

Descartes likened the transmission of light, by mechanical contact of particles, to a blind man sensing an object with his cane.

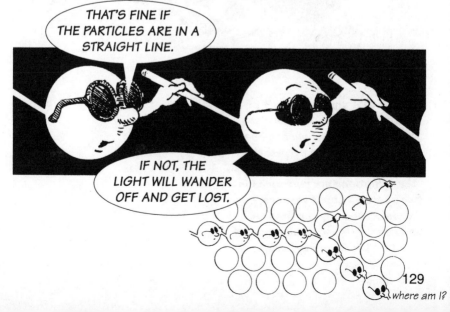

THAT'S FINE IF THE PARTICLES ARE IN A STRAIGHT LINE.

IF NOT, THE LIGHT WILL WANDER OFF AND GET LOST.

where am I?

Lost in Space

The behaviour of Descartes' vortex is found to violate Kepler's Laws, moreover it would slow down as its energy is "lost and swallowed up" in space.

THAT'S THE END OF ME!

"...so that the hypothesis of vortices is utterly irreconcilable with astronomical phenomena, and rather serve to perplex than explain the heavenly motions. How these motions are performed in free spaces without vortices, may be understood by the first book; and I shall now more fully treat of it in the following Book."

System of the World

"In the preceding books I have laid down the principles of philosophy; principles not philosophical but mathematical. It remains that, from the same principles, I now demonstrate the frame of the system of the world."

By observation of the four satellites of **Jupiter**, Newton finds that their periodic times are as the 3/2th power of their distances from its centre.

Jupiter's moons obey my Third Law

Kepler

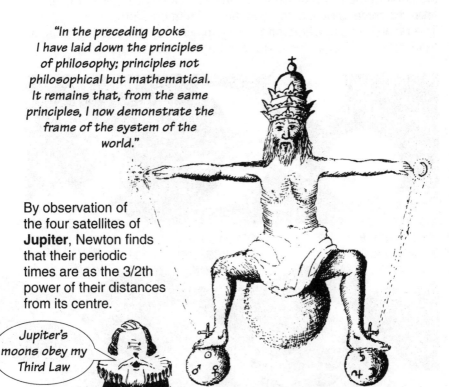

Drawing of Jupiter by Isaac Newton.

The satellites of **Saturn** are also found to obey Kepler's harmonic and area laws.

My Third Law applies to the five moons of Saturn, which I didn't even know existed.

Drawing of Saturn by Christiaan Huygens.

131

Various Phenomena...

Newton returns to his calculations of gravity on the Moon from the Plague Years. Then he had used an inaccurate value for the diameter of the Earth taken from Galileo. With a new value, he now finds the force to be exactly 1/3,600th of that on Earth.

The Moon's erratic wanderings are explained by the differing pulls of the Earth and Sun as the distances between them vary.

Newton considered the effects of three bodies (Sun, Moon, and Earth) on a ring of particles. With a ring of fluid equal to the radius of the Earth, he can calculate the tides.

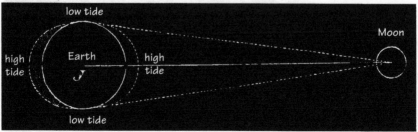

While Halley was on the island of Saint Helena observing the
southern skies, he noticed that his clock ran slow. Newton explained
that the Earth, because of forces generated by its rotation, was
squashed at the poles and bulged at the equator. The force of
Gravity on the pendulum of Halley's clock was less near the equator
than in London because he was 17 miles farther from the centre of
the Earth.

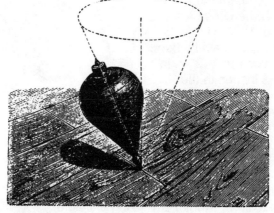

Since at least 129 BC, astronomers had known that the night sky, in
addition to its daily rotation, had a regular slow drift. This was called
the *Precession of the Equinoxes* and was unexplained.
Newton showed that it was due to the Earth's shape and the tilt of its
axis. There would be no effect on a perfectly spherical Earth. As it is,
the Sun's attraction tries to right the squashed Earth, while the Moon
tries to pull the bulging equator in line with its own orbit. The
combined effect on the Earth's tilted axis is to make it slowly rotate, a
complete circle taking 26,000 years.

The Tail

"And now we have described the system of the Sun, the Earth, Moon, and planets, it remains that we add something about comets."

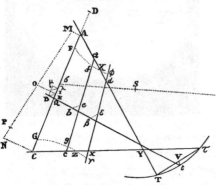

Newton devised a method of determining the path of a comet from only three observations. As an example he chose the great comet of 1680-81. He plotted the comet's path by hand with ruler and compasses — to a scale of 16.33 inches to the radius of the Earth's orbit. His drawing agrees with modern calculations to 0.0017 inch.

He found that the comet followed a parabola, (obeying Kepler's First Law) and swept out areas in proportion to time, (Kepler's Second Law).

I think I may infer that the tail is nothing else but a very fine vapour, which the head or nucleus emits by heat, becoming greater after it has passed by the neighbourhood of the Sun.

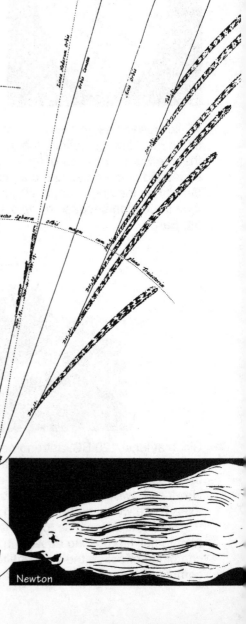

Newton

From the evidence of the phenomena, Newton concludes that the force of gravity exists, and that this self same force...

Causes bodies to fall to Earth.
Causes the Tides.
Holds the Moon in orbit round the Earth.
Holds satellites in orbit around their planets.
Holds the planets in orbit around the Sun.

It even steers the Comets which are only passing visitors, so it also applies outside the solar system

Everywhere the same laws hold, therefore this force is Universal Gravitation.

ALL BODIES WHATSOEVER ARE ENDOWED WITH A PRINCIPLE OF MUTUAL GRAVITATION. EVERY TWO BODIES GRAVITATE TOWARDS EACH OTHER IN PROPORTION TO THEIR MASSES AND IN INVERSE PROPORTION TO THEIR DISTANCES.

Thus dear reader while the Earth's gravitational field holds you stuck to your chair and weighs down the book in your hands, your own gravitational field attracts the book and the Earth, while the book itself...

The Book of Principles was writ in about 17 or 18 months, whereof about two months were taken up with journeys, & the MS was sent to ye Royal Society in spring 1686; & the shortness of the time in which I wrote it, makes me not ashamed of having committed some faults.

The reviews were glowing ...

This incomparable Author having at length been prevailed upon to appear in publick, has in this treatise given a most notable instance of the extent of the powers of the Mind: he seems to have exhausted his argument, and left little to be done by those that shall succeed him ... so many and so Valuable Philosophical Truths, as are herein discovered and put past dispute, were never yet owing to the Capacity and Industry of any one Man.

This one should be — Halley wrote it himself.

The World of Law

"No other work in the whole history of science equals the *Principia* either in originality and power of thought, or in the majesty of achievement. No other so transformed the structure of science, for the *Principia* had no precursor in its revelation of the depth of exact comprehension that was accessible through mathematical physics. No other approached its authority in vindicating the mechanistic view of nature, which has been so far extended and emulated in all other parts of science. There could be only one moment at which experiment and observation, the mechanical philosophy, and advanced mathematical methods could be brought together to yield a system of thought at once tightly consistent in itself and verifiable by every available empirical test. Order could be brought to celestial physics only once, and it was Newton who brought order. His is the world of law.**"**

— A. Rupert Hall, *From Galileo to Newton*.

Isaac can justly feel that the quarter-century he has spent cloistered in his ivory tower was well spent. At the age of 45, his place in history is assured. When he lifts his eyes from his papers he discovers there is more to life than books.

Fighting the Beast

In 1685 James II was proclaimed King. He, at once, set about tightening his grip on the country. As part of a programme to catholicize the universities he ordered a Benedictine monk, Father Francis, to be admitted to Cambridge.

Isaac had strong opinions about most things but there was nothing he hated more than monks.

Previously retiring, he came out and publicly defied King James. He was the spokesman for resistance at Cambridge.

He persisted to oppose the appointment on legal grounds despite dire warnings from the King and Judge "Bloody Assize" Jeffreys, who a year earlier ordered 300 hanged for rebellion.

King James II, 1633-1701

George Jeffreys of Wem, 1645-89

Newton put his career and even his life at risk. He was only saved from a sticky end by the landing of King Billy at Torbay.

Glorious Revolution

In 1688 the two traditionally antagonistic political parties, the *Whigs* and *Tories*, temporarily buried their differences to stop King James. They invited James' protestant son-in-law, a Dutch Prince, to invade England and turn out the King.

The term Whigs originally meant Scottish horse thieves!

And Tories were Irish footpads.

The bloodless *Revolution* gave Whigs control of the central state apparatus.

But the Tory squirearchy held on to local government in the country districts.

William of Orange, 1650-1702
King William III, 1689

TOGETHER THEY PUT ME ON THE ENGLISH THRONE — ALTHOUGH I ONLY SPEAK DUTCH!

Newton was rewarded for his anti-Catholic stance with a seat in the Parliament which decided the Revolutionary Settlement. He had an impeccable voting record, but spoke only once. Feeling a draught, he asked an usher to close the window.

He began spending most of his time in London. He lunched with the new King, and became acquainted with prominent figures like the philosopher John Locke, and enjoyed the admiration of a circle of young scientists.

Friends

Locke becomes an intimate friend. The two exchange views on subjects that Newton holds dear — science, economics, politics...

THE STATE OF NATURE HAS A LAW OF NATURE TO GOVERN IT.

MEN LIVING TOGETHER ACCORDING TO REASON, WITHOUT A COMMON SUPERIOR ON EARTH, WITH AUTHORITY TO JUDGE BETWEEN THEM IS THE STATE OF NATURE.

Newton

Locke

... but Newton also dares name, for the first time, views he has held in secret for 20 years. He speaks of his rejection of the Trinity (known as Arianism) and his discoveries of falsification in the Bible. Locke offers to help with publication of *An Historical Account of Two Notable Corruptions of Scripture*.

Newton decides to put out an edition in French in Holland, although public knowledge of his Arian beliefs would spell an instant end to his political and academic career, even force him into exile.

He takes the risk because he believes that a bill assuring complete religious toleration will be passed by Parliament. In the event, Catholics and Arians are excluded. Newton scrambles desperately to retrieve the manuscript from Holland where it is about to be published under his own name.

FIRST TIME I'VE BEEN PAID FOR NOT PUBLISHING A BOOK.

Jean Le Clerc

Newton's willingness to take such a risk in the first place speaks eloquently of the change that had taken place in him. Happy, confident, open, basking in his new-found fame.

140

One summer day, at a meeting of the Royal Society, Newton met a 25-year-old Swiss mathematician. Westfall says, "the attraction between the two was instantaneous".

Newton is the most honest man and the ablest mathematician that ever lived.

Nicholas Fatio de Duillier, 1664-1753

Fatio was a brilliant and devoted Cartesian. A mathematical groupie, he roamed Europe seeking out the famous. In 1689 he was visiting London with Huygens. By the autumn he was an ex-Cartesian.

For four years he and Newton were in regular correspondence. Fatio began to work on his own version of the Principles which he felt would be better than the original. Though Halley and Newton laughed at Fatio's extravagant ideas, he was indulged like a child.

Newton encouraged young mathematicians and scientists, he went out of his way to further their careers and even helped them with money. But there is no evidence of homosexual interest in Newton's life.

There were all sorts of practical reasons for Newton's accessibility to younger colleagues. He used them to copy and edit manuscripts, the chores Wickins had undertaken at Cambridge. They were used as messengers, translators, defenders in disputes and even to publish Newton's views under their own names. Not even Newton's enemies ever suggested anything worthy of comment.

NOTHING TO COMPLAIN ABOUT ON THAT SCORE.

Hooke

Then Fatio's enthusiasm abated, he admitted that he was abandoning mathematics. He had met a new friend, with whom he planned to set up a patent medicine business and make a fortune.

Friends noticed a sudden change in Newton...

141

Most Unfortunate

Pepys

HAS HE GONE MAD?

Sir,
I have neither ate nor slept well this twelve month nor have I my former consistency of mind. I must withdraw from your acquaintance and see neither you nor the rest of my friends any more.

Is. Newton

Locke

EMBROIL HIM WITH WOEMEN?

Sir,
Being of the opinion that you endeavoured to embroil me with woemen, when one told me that you were sickly and would not live I answered that 'twere better if you were dead ... I took you for a Hobbist.
your most humble & most unfortunate Servant
Is. Newton

Newton's breakdown has been blamed on various factors...

Clearly a case of mercury poisoning.

But he suffered no tremors or loss of teeth?

Or a fire caused by his dog Diamond.

He never had a dog.

Whatever the cause, the euphoria which followed the *Principia* slides into depression. Newton scuttles plans to to publish works on optics and mathematics, starts obstructing publication of the second edition of the *Principia*, and finally abandons alchemy in disillusionment. He leaves Cambridge for good to live in London.

Nobody Paying

A sympathetic Whig grandee and ex-pupil offers Newton a sinecure at the Royal Mint, a job, in principle, not requiring more attendance than he could spare.

I WOULD NOT SUFFER THE LAMP THAT GAVE SO MUCH LIGHT TO WANT OIL.

Charles Montague, Earl of Halifax, Chancellor of the Exchequer, 1661-1715

But at this point the British currency is in chaos. Over twenty percent of the coin is counterfeit. Only half the remainder has the correct weight because of the widespread criminal practice of "clipping" small pieces from the coins. Foreign countries are refusing to accept British coins.

MONEY STILL CONTINUING SCARCE. SUCH A SCARCITY THAT TUMULTS ARE EVERY DAY FEARED. NOBODY PAYING OR RECEIVING MONEY.

John Evelyn 1620-1706. A contemporary diarist.

The economic situation is so bad that the Treasury is on the verge of collapse, and threatening to take the Glorious Revolution with it. Which would bring yet another Restoration of the Stuart monarchy.

Threat to the State

No half measures for Newton, he throws himself into his new job with the same zeal he applies to everything he does.

HE PUTS 300 MEN AND 50 HORSES TO WORK 20 HOURS A DAY

PRODUCTION ROSE FROM £15,000 TO £120,000 A WEEK

NEWTON RECOINED SIX AND A HALF MILLION POUNDS IN THREE YEARS.

TWICE THE AMOUNT COINED IN THE PREVIOUS 30 YEARS!

THE GAME'S UP, GIBBONS!

Counterfeiting is such a threat to the survival of the State that it is decreed High Treason. With his life threatened, Newton relentlessly pursues the criminals, confronting them in person on their home ground in London's taverns and prisons. The underworld had never known such an organized assault.

Connivers at Abuses & Cheats

The most redoubtable of Newton's adversaries was William Chaloner.

Chaloner was the Moriarty of coiners, he manufactured above 30,000 guineas. He sent several innocent men to the gallows for reward. He escaped the same fate himself five times, twice by strangling witnesses. But after a career crowned only with success, he made the mistake of taking on Isaac Newton.

Chaloner publically accused the Mint of incompetence and offered a remedy. His plan was no less than to have his henchman, Holloway, appointed in charge of catching coiners, and himself made Supervisor of the Mint! Parliament was taken in and gave Chaloner permission to examine the Mint's secret machinery. Newton refused outright and set inexorably to work collecting the web of evidence which would ensnare Chaloner. After two years, Newton's dossier was as thoroughly researched as any proposition in the *Principia*. He ordered Chaloner's arrest. This time there was to be no escape.

Tyburn Tree

Justice which oft had been baffled b[y] him was now ready with her Iron hand[s] to break him to pieces

In 1697 there were 19 executions at Tyburn for crimes against the coinage. In all Newton sent 28 to the gallows. Was he a vindictive sadist?

The blood of the coiners and clippers nourished him.

Frank E. Manuel

Unable to slough the obligation off... it is not evident that Newton invested more concern in this aspect of his job than in others.

Richard S. Westfall

IF HE HAD LET THE MONEY AND THE GOVERNMENT ALONE & RETURNED TO HIS TRADE OF JAPANNING, CHALONER MIGHT STILL LIVE.

The President

Towards the turn of the century, Hooke fell ill and without his practical influence The Royal Society languished. It had fallen into a sorry state when compared to its heyday in the early 1670s.

FEW MEMBERS ATTEND THE MEETINGS

IT'S BECOME A CLUB FOR GOSSIP FOR BORED NATURAL PHILOSOPHERS!

Wren

Sloane

Hooke died in 1703, blind, isolated and rendered sordid by penury.
The pressure at the Mint had eased to such an extent that Newton found himself with time on his hands. He turned his attention to the Royal Society.

Elected President, a post he was to retain for the rest of his life, Newton set about revitalizing the society, but first he had a score to settle...

BURN HOOKE'S PORTRAIT!

He had a *Scheme for Establishing the Royal Society.*

NATURAL PHILOSOPHY CONSISTS IN DISCOVERING THE FRAME AND OPERATIONS OF NATURE, AND REDUCING THEM, AS FAR AS MAY BE, TO GENERAL RULES OR LAWS, — ESTABLISHING THESE RULES BY OBSERVATIONS AND EXPERIMENTS, AND THENCE DEDUCING THE CAUSES AND EFFECTS OF THINGS.

Whereas previous presidents had scarcely bothered to attend the weekly meetings, Newton missed only three over the next 20 years. To restore interest and increase membership he had a practical experiment performed at each meeting, starting with Francis Hauksbee's air-pump.

Officially recognized as the most eminent scientist in town, and with Hooke safely out of the way, Newton finally felt able to publish his Opticks.

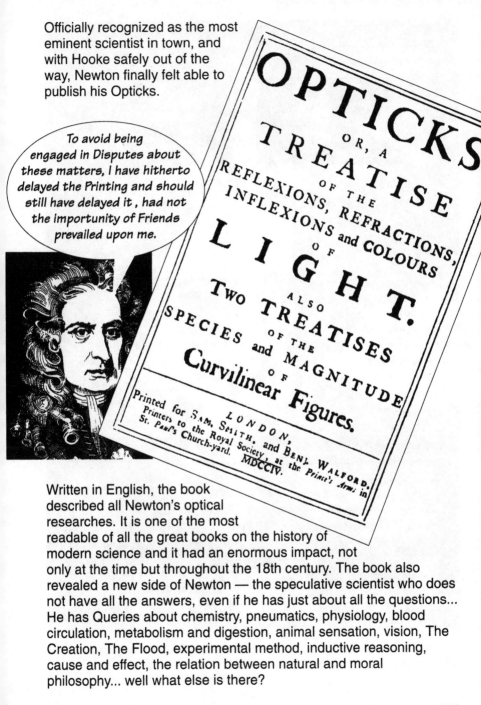

To avoid being engaged in Disputes about these matters, I have hitherto delayed the Printing and should still have delayed it, had not the importunity of Friends prevailed upon me.

OPTICKS

OR, A TREATISE

OF THE

REFLEXIONS, REFRACTIONS,

INFLEXIONS and COLOURS

OF

LIGHT.

ALSO

Two TREATISES

OF THE

SPECIES and MAGNITUDE

OF

Curvilinear Figures.

LONDON,

Printed for SAM. SMITH. and BENJ. WALFORD. Printers to the Royal Society; at the Prince's Arms in St. Paul's Church-yard. MDCCIV.

Written in English, the book described all Newton's optical researches. It is one of the most readable of all the great books on the history of modern science and it had an enormous impact, not only at the time but throughout the 18th century. The book also revealed a new side of Newton — the speculative scientist who does not have all the answers, even if he has just about all the questions... He has Queries about chemistry, pneumatics, physiology, blood circulation, metabolism and digestion, animal sensation, vision, The Creation, The Flood, experimental method, inductive reasoning, cause and effect, the relation between natural and moral philosophy... well what else is there?

God is Space

Newton had second thoughts about revealing his thinking about infinite space and recalled the book to cut out and replace the relevant page. But a few copies of the original edition escaped...

Newton's Headache

Newton desperately needed more accurate observations of the orbit of the Moon to perfect his Lunar Theory which was still incomplete...

IT'S THE ONLY THING THAT'S EVER GIVEN ME A HEADACHE.

Remedy for Headache By Isc. Newton Esq. Bind your head with a garter till the crown is numbed. This cools the head by retarding the circulation of the blood. Q.E.D.

There was an Astronomer Royal who was charged with calculating the positions of the moon and stars, but he had produced nothing for 30 years.

John Flamsteed, the first Astronomer Royal, began making observations, from the new observatory at Greenwich, in 1676.

Money for the building was raised by selling old and decayed gunpowder.

Flamsteed, crippled by gout, irritable and defensive, refused to part with his observations on the grounds that they were his private property.

I USED MY OWN INSTRUMENTS!

YOUR RESULTS ARE CROWN PROPERTY! AND SAY SIR WHEN YOU TALK TO ME!

Flamsteed

Sir Isaac, knighted 1705, the first scientist to be so honoured.

Gold Ring

Newton paid Flamsteed £180 to calculate positions of the Moon. But Flamsteed spent the money on the fixed stars which were of no use to Newton. When Flamsteed was eventually forced to part with his observations, Newton found them riddled with mistakes!

HOW COULD YOU IMAGINE THESE ERRORS WERE INTENDED?

Flamsteed took every opportunity to belittle Newton...

Well I never!

Doing good has always been a rule of my life, but not of his. Newton is a great rascal, he has stole two stars from me!

Thos. Weston   Flamsteed

...or to obstruct him, or accuse him of wrong doing.

NEWTON IS A GREAT RASCAL. HE WORKED WITH THE ORE I HAD DUG.

IF THAT DOG FLAMSTEED DUG THE ORE, I MADE THE GOLD RING.

Newton

Furious, Newton went through the *Principia* and struck out every reference to Flamsteed, and for good measure had him thrown out of the Royal Society for non-payment of dues.

Sextant

Regiomontanus, 1436-76

Regiomontanus was first to suggest using the Moon's position to determine the position of ships at sea. But the problem of how to measure accurately remained.

In 1707, through bad navigation the British fleet under Admiral Shovel struck the rocks of the Scilly Isles. Two thousand men, including the admiral, and a treasure were lost. This brought the problem to the attention of the authorities.

Newton was aware that an accurate clock would enable position to be found, but as he put it, "When the longitude at sea is once lost it cannot be found by any watch".

The lunar distance method required a navigator to measure the distance between the Moon and a fixed star and the horizon to better than 2 minutes of arc. Newton himself had made a double reflecting sextant which enabled sighting on the moon to that degree of accuracy, even on the rolling deck of a ship. He showed it to the Royal Society in 1699, but dropped it after Hooke claimed to have made one thirty years before.

In 1714 Sir Isaac was appointed by Parliament to lead a committee seeking the best way to "Discover Longitude at Sea". A reward of £20,000 was offered. Newton spent the following years dealing with a flood of cranky proposals.

The prize eventually went to the clockmaker John Harrison for his chronometer which sailed with Captain Cook.
(The sextant was re-invented independently in America and England in 1731)

153

Chagrin and Mortal Anger

The conflict with Flamsteed was a minor irritant compared with the feud between Newton and Leibniz, a violent dispute that ran for two decades and continued beyond the German philosopher's grave.

It began in 1684, when Leibniz published his discovery of the calculus without mentioning Newton's progress on the subject.

Leibniz, one of the supreme intellects of all time, but as a human being he was not admirable

Bert Russell

Newton might have let it pass had not Leibniz gone on to claim that he was the sole inventor of the calculus. Newton's friends were up in arms...

Leibniz stole the calculus from Newton.

Isaac, your notions are passing abroad under Leibniz's name!

John Wallis

...and finally a furious Newton allowed the full text of the letters between them from 1676 to be published — but this only drove Leibniz to deny everything!

The German, while praising Newton in public, began vicious anonymous attacks in scientific journals . Newton replied in kind...

Great men are like women who never give up their lovers except with the utmost chagrin and mortal anger. And that, gentlemen is where your opinions have got you.

Caroline of Ansbach, 1683-1737. Queen Consort.

Empirical

The row did not end with Leibniz's death. It emerged that he had been shown Newton's *fluxions* by Collins, the publisher, in London in 1676 — something Newton had begun to suspect.

YOU CRIMINAL!

WHAT ABOUT YOU WITHHOLDING CALCULUS FROM THE WORLD FOR 30 YEARS!

Each of the contestants rallied supporters and the dispute raged on into the 19th century, causing a split that would shut English mathematicians off from the mainstream of continental progress for a hundred years and deny them use of something that is beyond dispute — Leibniz's superior system of notation. The German's dx, dy and $\int$, which we use today, only reached England a hundred years later.

The priority dispute was not the only point of contention between these two intellectual giants. Newton was a figurehead of the philosophical movement called English Empiricism.

Newton draws comparatively modest conclusions from a broad survey of facts...

Whereas Leibniz builds a vast edifice of deduction on a pin-point of logical principle.

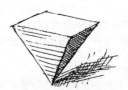

He can correct a flaw without causing total collapse...

Leibniz's structure is unstable and the slightest flaw brings it down in ruins.

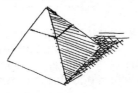

Chronology

On top of the disputes, Newton was obliged to spend endless time during his final years revising his theological work in order to cover up his Arianism! Princess Caroline had got wind of his preoccupation with religious history and asked to see a copy of his writings. Threatened with ruin if the truth came out, Newton produced an extract from his work leaving out all hints of his anti-Trinitarianism. But when a pirated edition of the extract appeared in Paris it provoked a wave of criticism, demanding a response from Newton. He was forced to re-write the whole thing.

With the Arianism removed, what is left?

Newton's startling innovation was **astronomical dating**. He compared descriptions of the night sky in ancient literature and using the *precession of the equinoxes* calculated dates from the positions of the stars, shortening accepted chronologies by 500 years. He put the voyage of the Argonauts in 937 B.C.

NEWTON'S GOAL WAS TO PROVE THAT THE JEWISH CIVILIZATION WAS OLDER THAN THE GREEK.

It was a bomb-shell in the staid world of the ancient historians. Voltaire wondered if the world of science might find it too much to grant to one man the honour of having perfected physics, mathematics and history at the same time...

"It would be a kind of universal sovereignty difficult for personal pride to accept!"

Voltaire

THE

CHRONOLOGY

OF

ANCIENT KINGDOMS

AMENDED.

To which is Prefix'd,

A SHORT CHRONICLE *from the First Memory of Things in* Europe, *to the Conquest of* Persia *by* Alexander *the Great.*

By Sir *ISAAC NEWTON.*

LONDON:

Printed for J. TONSON in the *Strand,* and J. OSBORN and T. LONGMAN in *Pater-noster Row.*
MDCCXXVIII.

Newton did not live to see the publication of the doctored full-scale version, *Chronology of Ancient Kingdoms Amended.*

Nor did he outlive publication of his *Observations upon the Prophecies* — a subject that had engaged him since the 1670s.

OBSERVATIONS
UPON THE
PROPHECIES
OF
D A N I E L,
AND THE
APOCALYPSE
OF
St. ℐ O H N.

In Two Parts.

By Sir *ISAAC NEWTON.*

LONDON,
Printed by J. Darby and T. Browne in *Bartholomew-Close.*
And Sold by J. Roberts in *Warwick-lane,* J. Tonson in the Strand, W. Innys and R. Manby at the West End of St. Paul's Church-Yard, J. Osborn and T. Longman in *Pater-Noster-Row,* J. Noon near *Mercers Chapel* in *Cheapside,* T. Hatchett at the *Royal Exchange,* S. Harding in St. Martin's lane, J. Stagg in *Westminster-Hall,* J. Parker in *Pall-mall,* and J. Brindley in *New Bond-street.*
M.DCC.XXXIII.

He was in no hurry to publish because the end of the world was not imminent.

NEWTON SAYS THE WORLD WILL END IN THE YEAR 2132.

Early in 1725 Newton moved to Kensington for his health.

"Sir Isaac lives a little way out in the country. He is extremely kind and serviceable, but he is much failed and not able to do as he has done." — Conduitt

On an attempt to persuade him to ride to church instead of walking...

USE LEGS & HAVE LEGS.

He began to give away money, and burned some papers. In March his distemper was diagnosed as a stone in the bladder.

The End

At last on his death bed Newton declared the belief that he had held secretly for fifty years, he refused the last sacrament. He became insensible on Sunday March the 19th, and died the next morning at one o'clock.

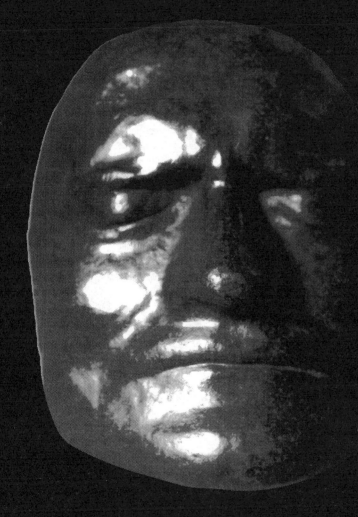

"The pain rose to such a great height
that the bed under him, and the very room
shook with his agonys, to the wonder of those
present. Such a struggle had his great soul to
quit its earthly tabernacle."

Letter on England

Newton was gone but he left a permanent stamp on English society. Everyone there saw the world through Newtonian eyes; quite differently from the continent where Descartes still held sway.

Voltaire

The French philosopher exiled in England at the time of Newton's funeral, was impressed by the intellectual climate there. He described various facets of English life and beliefs in his book, *Lettres Philosophiques*. He depicts England as the land of freedom, tolerance and progress in contrast to the superstitious feudal tyranny across the channel. Published in France in 1734 it was immediately banned.

In its impact on social, economic and political thinking, Voltaire's *Letters on England*, turned out to be one of the most influential books of the century. It formed an essential ingredient in the explosive mixture brewing in France.

Newton's scientific triumphs had paved the way for Locke's democratic philosophy, which in turn would help to spark revolutions across the world, starting with America in 1776, and would end up being written into almost all the constitutions of modern times.

"In an age which produces the incomparable Mr. Newton, it is ambition enough to be employed as an under-labourer in clearing the ground a little and removing some of the rubbish that lies there in the way to knowledge."
— *John Locke*

Rights of Man

Locke

A GOVERNMENT IS NOT FREE TO DO AS IT PLEASES. THE LAW OF NATURE, AS REVEALED BY NEWTON, STANDS AS AN ETERNAL RULE TO ALL MEN.

Liberalism conceived **The Inalienable Rights of Man**. In 1776 these were written into the Declaration of Independence to launch the American Revolution.

This is from Locke's Treatise on Government.

This is Euclid.

We hold these truths to be...

...self evident...

...that all men are equal, that they are endowed by their creator with certain inalienable rights...

Jefferson

Franklin

160

Revolting French

The democratic ideal of equality proved to be the sticking point for Louis XVI, which provoked the French Revolution of 1789.

You can have a free press, free trade, tax reform and land reform but I'm not equal and that's all there is to it.

Louis XVI, 1754-93

The French people insisted that they too were born equal and had inalienable rights. *The Declaration of the Rights of Man and of the Citizen* was adopted in 1789. The Pope condemned it as impious.

"I turned my eyes to the Schools and universities of Europe, and there behold the Loom of Locke, whose woof rages dire Washed by the Water-wheels of Newton..."

William Blake

Locke. that sower of evil seed, which quickened in the warm mud of Paris, produced the revolutionary monster that has devoured Europe.

161

Republican Virtue

According to J.K. Galbraith, the purchase of a coat in the 18th century by an average citizen was comparable to buying a car, or even a house today. In this context the mass production of cheap cloth was democratic and revolutionary. This was the spirit of the new technology that was to be put to the service of the American Revolution.

Nothing is good or beautiful, but in the measure that it is useful. The invention of a machine, or the improvement of an implement is of more importance than a masterpiece of Raphael.

Architecture, Sculpture and pictures have conspired against the rights of mankind.

John Adams, 1735-1826.
Second American President.

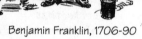

Benjamin Franklin, 1706-90

Opinions were divided as to the effects which the application of science, through industrialization, might have have on the virtue of the new-born Republican society.

LET OUR WORKSHOPS REMAIN IN EUROPE. THE MOBS OF THE GREAT CITIES ADD JUST SO MUCH TO THE SUPPORT OF PURE GOVERNMENT AS DO THE SORES TO THE STRENGTH OF THE HUMAN BODY.

Thomas Jefferson, 1743-1826.
Third American President.

Nonsense on Stilts

Locke and Newton may have been the secular patron saints of the political revolutions, but the technological revolution which followed forgot Newton's respect for Nature. The new industrialists cast around after more profitable ideas.

RIGHTS OF MAN, NONSENSE; IMPRESCRIBABLE RIGHTS OF MAN, NONSENSE ON STILTS.

At last someone with a bit of sense.

Hobbes

Jeremy Bentham, 1748-1832.
The Auto Icon, his mummified body, is kept in a Cupboard at University College London.

Arithmetic not calculus was the fundamental tool of the Industrial Revolution. And if profit and loss were applied to morals and politics, happiness and pain could be totted up. Subtract pain from gross happiness, result net happiness.

Hobbesian individualism, man unlimited in his appetites, a constant state of war, competition running free in economics and culture, were the order of the day. Moreover the side-effects of this unrestrained struggle were good for the entire race.

MILLIONAIRES ARE A PRODUCT OF NATURAL SELECTION.

William Graham Sumner

Celestial Revolutions

So much for earthly revolutions. Sir Isaac's theories were being severely tested in the heavens. There remained celestial irregularities that he had been unable to account for, thereby providing his opponents with ammunition. And as for "action at a distance"...

"This is, in effect, to return to occult qualities, and what is even worse, inexplicable ones. One would renounce Philosophy and Reason, opening an asylum for ignorance and laziness."

"The Principle of Attraction seems to me absurd. How could he give himself all the trouble of making such a number of investigations and difficult calculations that have no other foundation."

Leibniz

Huygens

Under Voltaire's influence, the scientists of the Enlightenment in France took up Newton and followed his theories to their proper conclusions.

"IF SOMEONE WERE ABSURD ENOUGH TO BELIEVE IN SCREW-FORMED MATTER WE WOULD SAY THIS MAN IS A CARTESIAN; IF HE SHOULD BELIEVE IN MONADS, HE IS A LEIBNIZIAN. BUT THERE ARE NO NEWTONIANS. IT IS THE PRIVILEGE OF ERROR TO GIVE ITS NAME TO A SECT."

Gabrielle-Emile, Marquise de Châtelet, 1706-49. Translator of *Principia* into French.

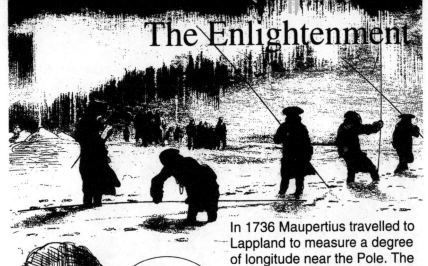

The Enlightenment

In 1736 Maupertius travelled to Lappland to measure a degree of longitude near the Pole. The result confirmed Newton's predictions. A degree of longitude was indeed longer at Torneå. The Earth was flattened near the poles.

In Lappland one degree is 57,395 toises, in France 57,060.

Pierre Louis Moreau de Maupertius,
1689-1759

The mathematician Lagrange added the dimension of time to Cartesian three dimensional space, creating four dimensional space-time. At the age of 28 he solved the *Three Body Problem*, proving that the Moon does indeed follow Newtonian Law.

Laplace in his *Mécanique Celeste* answered the question, Is the Solar system stable? In spite of all the perturbations of the planets each upon the other he showed that it is.

A century after Newton's death, Laplace said this of his laws:

EACH NEW DIFFICULTY WHICH HAS ARISEN HAS BECOME FOR IT A SUBJECT OF TRIUMPH - A CIRCUMSTANCE WHICH IS THE SUREST CHARACTERISTIC OF THE TRUE SYSTEM OF NATURE.

Pierre Simon de Laplace, 1749-1827

Eppur Si Muove*

Inside the Pantheon in Paris, in 1871, Foucault finally succeeded where others had failed. He demonstrated conclusively that the Earth is moving. He hung a 60 kg. copper ball on a 68 metre wire from the top of the dome. When set swinging a needle on the ball traced a line in sand spread out on the floor.

A PENDULUM! WHY DIDN'T I THINK OF THAT?

Galileo

Lines traced by subsequent swings showed that the pendulum was drifting. After five minutes the distance between the start of the first and last lines was several centimetres. In fact, the pendulum stayed swinging in the same place, but the Earth was turning under it, from West to East.

*"But still, it **does** move." The words Galileo didn't speak at his trial.

The Limit

On the mathematical front, there was continued carping.

What are these fluxions? The velocities of evanescent increments. And what are these same evanescent increments? They are neither finite quantities, nor quantities infinitely small, nor yet nothing. May we not call them the ghosts of departed quantities?

Bishop Berkeley, 1685-1753.

Zeno's Achilles paradox remained a sticking point. "Does a variable which approaches a limit ever attain it?"
In 1872, Weierstrass finally exorcised Zeno by treating calculus as number alone, excluding geometry completely. The idea of curves being generated by points in motion was wrong. Weierstrass said that the limit does not involve the idea of approach at all but is static. But, just when the calculus was, at last, established on a rigorous logical basis, the world of geometry was sent reeling.

Euclid's Achilles heel was the parallel postulate. He never did prove that parallel lines never meet. Other *non-Euclidean* geometries could be constructed.

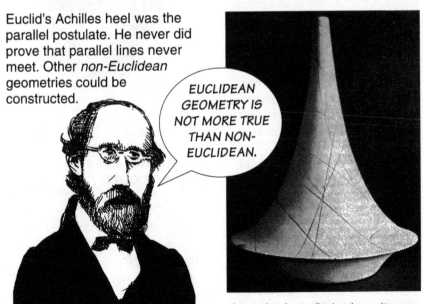

EUCLIDEAN GEOMETRY IS NOT MORE TRUE THAN NON-EUCLIDEAN.

Bernhard Riemann, 1826-1866

A pseudosphere, all triangles on its surface follow consistent laws.

Lost Planet

In 1781, William Herschel discovered a new planet, Uranus. But as its orbit was plotted over the following years, the mathematician Bouvard was baffled by its path. He concluded that Newton's Laws did not apply so far from the Sun.

U. J. J. Le Verrier didn't loose faith.

I PREDICT ON THE BASIS OF NEWTON THAT THE DISTURBANCE IS CAUSED BY AN UNKNOWN PLANET.

Le Verrier calculated the position of the unknown object. Galle searched the sky where Le Verrier said it should lie. In 1846 he found Neptune.

The discovery of Neptune was perhaps the most spectacular confirmation of Newton's Laws, almost 200 years after Isaac began calculating the movements of the heavens, in his mother's apple orchard at Woolsthorpe.

Urbain Le Verrier, 1811-1877

I'M AFRAID THE TRUTH IS SLIGHTLY MORE COMPLICATED.

Emboldened by this success, Le Verrier predicted yet another planet to account for the slight inexplicable drift in Mercury's obit. He called this new planet Vulcan.

Einstein

In trying to account for the creeping of Mercury's orbit, Einstein considered why bodies of different weights fall at the same speed. He felt that gravitation must depend on the structure of space and time. Striving to calculate the properties of space-time, he found that the geometry already existed, created by Riemann.

The Moon is kept in its orbit around the Earth by curvature of space-time.

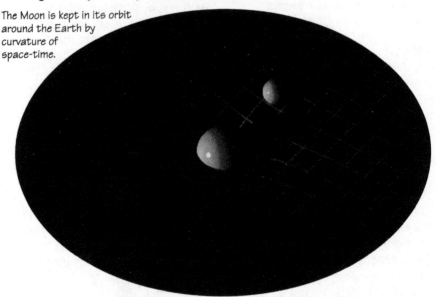

To test his theory, Einstein predicted that the path of a ray of light should be bent by the effect of a gravitational field. A measurable effect ought to be observable during an eclipse of the Sun. In 1919 the Royal Society sent an expedition to the island of Principe off the West-coast of Africa. Eddington found that light did indeed bend! Newton's fixed absolute space had ceased to be.

NEWTON, FORGIVE ME. YOU FOUND THE ONLY WAY THAT WAS POSSIBLE FOR A MAN OF THE HIGHEST POWERS OF INTELLECT AND CREATIVITY. THE CONCEPTS THAT YOU CREATED STILL DOMINATE THE WAY WE THINK IN PHYSICS.

Dual Nature

While Einstein put paid to Newton's concept of absolute space, he reinstated his theories on the duality of light and the existence of atoms, both of which had been discredited.

At the end of the 19th century, no one believed that atoms existed.

ATOMS ARE THE FIGMENT OF A WEAK IMAGINATION.

Leibniz

Newton sometimes treated light as a wave, sometimes as a particle. After being regarded for generations as an artificial attempt to save a dying theory, this guess of Newton has been proved to be a supreme example of the intuition of genius. With his paper on the photo-electric effect, Einstein showed that Light is both a wave and a particle at the same time.

Boltzmann, the last of the old school, was driven to suicide by the derision of the modern young thinkers, followers of Ernst Mach.
With his paper on Brownian movement, Einstein gave concrete evidence of the existence of atoms.

ARE NOT GROSS BODIES AND LIGHT NOT CONVERTIBLE INTO ONE ANOTHER?

That's like, $E=mc^2$

Faced with this dual nature of matter, a new kind of mechanics was invoked. In **Quantum Mechanics**, matter interacts and forces are transmitted by tossing **quanta** (packets of energy) back and forth. As Newton had been led to suspect by alchemy, energy and matter can be converted into each other and the quantum of light is the **Photon.** There are generally agreed to be four different forces in nature. **Electro-magnetism**, which underlies all chemistry; its quantum is the **Photon.** The **Strong Force** binds the atom together; its quantum is the **Gluon.** The **Weak Force** causes radioactive decay; its quantum is the **Boson.** The weakest of all **Gravity** has the weightless **Graviton.**
There have been various unsuccessful attempts to integrate these disparate forces into one grand **Theory of Everything** (**TOE**). Einstein spent the latter half of his life trying to reconcile quantum mechanics with gravity into a Unified Field Theory.

At a Distance

Are we any nearer understanding what Gravity is?

If there are **Gravitons** there must also be **Gravity Waves**. Extremely sensitive detectors have been built, but as yet no gravity wave has been recorded. Gravity resists quantization and is still only defined as a consequence of the curvature of space-time.

On the other hand, by measuring gravity in an Australian mine, Ephraim Fischbach has come to the conclusion that there is a fifth force in nature called **Hypercharge**. This anti-gravitational force would make itself felt up to 600 feet distant and its effect would be to make a feather fall faster than a brick in a vacuum!

A Superstring, much enlarged.

Green   Schwarz

String theory explains forces not as interacting point-like particles but as infinitesimally small, winding, curling, one-dimensional strings.

But whatever the future may bring...

LET NO ONE SUPPOSE, HOWEVER, THAT THE MIGHTY WORK OF NEWTON CAN EASILY BE SUPERSEDED BY RELATIVITY OR ANY OTHER THEORY. HIS GREAT AND LUCID IDEAS WILL RETAIN THEIR UNIQUE SIGNIFICANCE FOR ALL TIME AS THE FOUNDATION OF OUR WHOLE MODERN CONCEPTUAL STRUCTURE IN THE SPHERE OF NATURAL PHILOSOPHY.

Einstein

A Piercing Eye

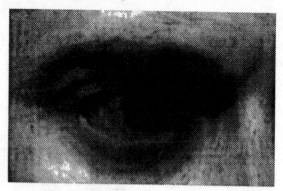

One reason for a widespread misapprehension of Newton's character may be that the portraits of him in circulation are stiff and inhuman with the air of a dead fish, or at best a virtuous dotty sheep; he who is described as "Happy and vigorous, with a lively and piercing eye."

There are very good portraits to see, all by Godfrey Kneller. One, done when Newton was 45, about the time of the Principia, belongs to the Portsmouth estate. Another very good one from 1702 is in The National Portrait gallery. Even Kneller's portraits of Newton in old age are filled with energy, there is a great one in Warsaw.

A mediocre painter can only create a misleading portrait. Frank Manuel draws some pretty profound conclusions about Newton's character and health from paintings by Vanderbank when it is the state of British portrait painting that was in poor health. Avoid anything by Vanderbank, Seeman and Thornhill, but also look out for copies. The Queen herself owns a lifeless copy of Kneller's masterpiece of 1689.

Thanks to Stephen Croall without whose contribution this book would never have got started. If there are any jokes they are probably his, the boring bits I did on my own. Thanks to Bill Brown for insisting that I read The Sleepwalkers, and thanks to the staff of Réanimation P. Even at Hôpital Laennec for saving the life of my wife, otherwise the book would never have been finished.

— Paris, Spring 1993

Books

For a general look at the history of man's changing view of the universe before Newton, you can do no better than to read **The Sleepwalkers** by *Arthur Koestler*. First published by Hutchinson in 1959 and available in paperback from Penguin.

The standard biography of Newton is **Never at Rest** by *Richard S. Westfall*, it appeared in 1980 from Cambridge University Press, and is unlikely to be surpassed.

There is a different sort of biography, **A Portrait of Isaac Newton** by *Frank E. Manuel* published by Muller. If you think Newton found the laws of gravity because he was powerfully drawn to distant persons, his dead father and his re-married absent mother, then this is the book for you. Full of fascinating background.

Newton's own version is available in the **Principia**, the Motte translation revised by Cajori, two volumes from the University of California Press.

His more readable **Opticks** is available from Dover Books, as are *Galileo's* **Two New Sciences** and *Descartes'* **Geometry**.

I. B. Cohen has has edited **Isaac Newton's Papers & Letters on Natural Philosophy** published by Harvard University Press. This is a collection of all of *Newton's* other published texts that were read during his lifetime. Produced in facsimile they really give a taste of what it was like to meet these ideas at the time.

For information on Neo-Platonism and how it affected the birth of modern thought **The Foundations of Newton's Alchemy** by *Betty Jo Teeter Dobbs* is available from Cambridge University Press, as is **From Paracelsus to Newton, Magic and the making of Modern Science** by *Charles Webster*.

If you just want the facts, **The Newton Handbook** by *Derek Gjertsen*, published by Routledge & Kegan Paul, puts them at your fingertips. A reference book, with five hundred entries in alphabetical order covering every aspect of the Newton's life and work. Well-written invaluable, a must for all Newtonians.

Journey Through Genius by *William Dunham* (Penguin) doesn't just walk you through Newton's discovery of the Binomial Theorem, but explains all the great theorems of mathematics from Hippocrates to Cantor and the Transfinite Realm.

Sir Thomas Heath's indispensable two volume **A History of Greek Mathematics** (1921) is available from Dover as is **Euclid's Elements** with commentary by *Sir Thomas*. Also to be recommended from Dover **The History of the Calculus and its Conceptual Development** by *Carl B. Boyer* and **The Origins of the Infinitesimal Calculus** by *Margaret E. Baron* with particularly fine diagrams.

Why a Dog?

What's the point of drawing Kepler like a dog apart from the fact that he often thought he was one?
The two portraits of Kepler below are typical, but the more you know about his life the less likely his hairstyle becomes.

Before the advent of photo-reproduction, portrait paintings were largely known through engraved copies which could be printed and widely distributed. These images take on a life of their own as they spawn ever less accurate copies of themselves. The starting point here is a portrait which never looked like Kepler in the first place and it has metamorphosed as fashions change as to what a German astronomer ought to look like.

Stamps of Einstein exhibit a similar relativistic effect as he seems to take on the racial characteristics of whatever country issues the stamp.

WOOF, WOOF!

On the left is a rather more likely portrait of Kepler; dog-like is at least a true picture of his character.